Facility Location and the Theory of Production

Facility Location and the Theory of Production

by

Arthur P. Hurter, Jr.

Department of Industrial Engineering and Management Sciences
Northwestern University

and

Joseph S. Martinich

School of Business Administration
University of Missouri – St. Louis

Kluwer Academic Publishers
Boston / Dordrecht / London

Distributors for North America:
Kluwer Academic Publishers
101 Philip Drive
Assinippi Park
Norwell, Massachusetts 02061 USA

Distributors for the UK and Ireland:
Kluwer Academic Publishers
MTP Press Limited
Falcon House, Queen Square
Lancaster LA1 1RN, UNITED KINGDOM

Distributors for all other countries:
Kluwer Academic Publishers Group,
Distribution Centre,
Post Office Box 322
3300 AH Dordrecht, The NETHERLANDS

Library of Congress Cataloging-in-Publication Data

Hurter, Jr., Arthur P.
Facility location and the theory of production.
Includes index.
1. Industry—Location—Mathematical models.
2. Production functions (Economic theory)
I. Martinich, Joseph Stanislaus, 1950–
II. Title.
HD58.H87 1989 658.2'1 88-13196
ISBN 0-89838-283-1

Printed in the United States of America.

To Florence, Vicki, and Michael for making it all worthwhile.

Contents

Preface

The design and location of production facilities are important aspects of corporate strategy which can have a significant impact on the socio-economy of nations and regions. Here, these decisions are recognized as being interrelated; that is, the optimal plant design (input mix and output level) depends on the location of the plant, and the optimal location of the plant depends on the design of the plant. Until the late 1950s, however, the questions of where a firm should locate its plant and what should be its planned input mix and output level were treated, for the most part, as separate questions, and were investigated by different groups of researchers. Although there was some recognition that these questions are interrelated [e.g., Predohl 1928; Hoover 1948; Isard 1956], no detailed analysis or formal structure was developed combining these two problems until the work of Moses [1958]. In recent years scholarly interest in the integrated production/locaton decision has been increasing rapidly.

At the same time that research on the integrated production/location problem was expanding, significant related work was occurring in the fields of operations research, transportation science, industrial engineering, economics, and geography. Unfortunately, the regional scientists working on the production/location problem had little contact with researchers in other fields. They generally publish in different journals and attend different professional meetings. Consequently, little of the recent work in these fields has made its way into the production/location research and vice versa.

The primary purpose of this book is to bring together in a single coherent volume all of the work previously published on the integrated production/location problem. This book is intended to provide a complete and rigorous treatment of the topic; proofs are given for major results and references are provided for relevant details. At the same time, in order to develop an understanding of the current state of the art in production/location research, it is necessary to explore how this work contributes to and draws upon work in other disciplines and other problems. This exploration is pursued throughout the book.

In addition to providing a unified presentation on production/location theory and its relationship to work in other disciplines, this volume makes several other specific contributions.

1. Given the prominent role of both national and local governments in the market economy, models that facilitate the evaluation of government policies in terms of their effects on the economic units of the economy are increasingly necessary. For many years economists have been studying the effects of public policies such as taxation, incentives, and regulations on the economic behavior of the firm. However, they have paid little attention to the locational or spatial aspects, especially when the policies are administered in a spatially uniform manner. Many government policies, even those that are not primarily meant to be economic (e.g., pollution regulations), affect both the production decisions (e.g., input mix) and location decisions of firms. In fact, the effect of public policies on a firm's plant location decision is often *through* their effect on the production process. Consequently, an integrated model of the firm's production and location decisions provides a better framework for analyzing the (sometimes subtle) effects of government policy. The integrated production/location model provides a format for simultaneously evaluating the technological, production, and locational effects of taxes, incentives, and regulations. In recent articles [e.g., Martinich and Hurter 1985] we have illustrated the policy analysis potential of these models, especially stochastic forms of the model. That work is expanded here, and the general issues of using taxes and incentives to influence locational decisions is examined in light of the model. We devote two full chapters to this topic.

2. Production/location research has been criticized for its abstract, theoretical nature. The resulting models exhibit a degree of mathematical complexity which makes it difficult to obtain mathematically precise results that are both instructive and general. The complexity, abstractness, and mathematical sophistication employed has undoubtedly limited access to the more recent work on production/location problems, and limited its application in fields such as public policy. We have attempted to improve the accessibility and clarity of the models by including many numerical examples to motivate and illustrate the theoretical results. These numerical examples make it possible for readers to understand the essence of the

theorems without mastering the mathematical details. In addition, numerical illustrations are used to demonstrate conjectured relationships that are not readily provable because of the mathematical complexity of the problem. This approach may not be satisfying to mathematical purists, but it greatly expands the insight that can be drawn from these models and increases their potential application.

3. Parametric uncertainty and risk preferences were first incorporated into the production/location problem only recently [Mai 1981; Martinich and Hurter 1982]. This book includes extensive discussion of the published work on the stochastic production/location problem, and presents new, previously unpublished work. The addition of uncertainty makes the problem more realistic (and complex), and takes advantage of recent research on economics under uncertainty.

4. In order to develop an understanding of the interaction among the producton decision, the location decision, and the problem parameters, extensive sensitivity analysis is necessary. Very little sensitivity analysis has appeared in the literature, so we have devoted considerable space to this topic.

5. The published work on production/location theory has dealt almost exclusively with one-facility models because the mathematical complexity of the production/location problem often makes multifacility models analytically intractable. Multifacility models are presented in this book, however, to highlight their computational complexity; to compare them with other more common location models; and to show under what conditions the problems may be tractable, at least with approximation methods.

6. Little attention has been given in the literature to the solution aspects of production/location problems. For each of the major categories of models, solution procedures are presented; and, where appropriate (e.g., multifacility problems), they are compared to algorithms that have been developed for other location problems, such as the Steiner-Weber problem and the fixed-charge problem. Some new algorithmic results are presented along with numerical illustrations.

Our overall goal in writing this book was to provide an integrated presentation of the state-of-the-art in production/location research. Further, we have attempted to present the material in such a manner that it is of interest and accessible to potential readers with a variety of backgrounds. A knowledge of microeconomics and calculus, and in some cases, probability theory, is necessary to follow the formal proofs and the more technical results. Nevertheless, we believe that our exposition and numerical examples should make the essential features and potential applications of the theory understandable to those without such a background. Our approach makes the book usable for special courses that cover production/location problems, as well as a research reference for regional scientists, economists, industrial engineers, operations researchers, and policy analysts.

We would like to acknowledge the support of The Technological Institute of Northwestern University, the School of Business Administration and the Office of Research of the University of Missouri–St. Louis, the School of Business Administration of Washington University, and the National Science Foundation, which supported much of the underlying research on which this volume is based. Special thanks for invaluable assistance in preparing the manuscript are due to June Wayne, George Mach, and Mary O'Brien.

Facility Location and the Theory of Production

1 INTRODUCTION

The construction of a new production facility by a firm requires that the firm make several interrelated decisions: how large should the plant be (planned output rate), what production method should be used (input mix), and where should the facility be located? In the traditional literature, and often in practice, these questions are addressed separately. The size and production method of the plant are usually determined by engineers, on the basis of sales projections, assumptions about the prices and availability of resources, and knowledge of the available technologies. The location of the plant is often selected by a separate team which may be made up of location specialists or a mix of people from industrial engineering, marketing, planning, and transportation, who assume a given plant size and design. This separation of the production decision from the location decision has been carried over into academic research. Microeconomists have focused on the firm's input mix and output decisions with minimal regard for spatial considerations, while location theorists, especially those in operations research, management science, and transportation science, have focused on where to locate plants for given production levels or techniques (cost functions). The fundamental tenet of this book, and integrated production/location theory in general, is that if a range of plant sizes, production techniques and locations are available, then the optimal size, input mix, and location of the facility are interrelated.

The Structure of Production/Location Problems

In order to make the types of problems to be considered in this book more tangible, we will begin by looking at a simplified, yet realistic, example. Suppose a firm wishes to establish a paperboard plant to produce a grade of paperboard that customers will use to manufacture folding cartons, fast-food trays, etc. The firm wishes to locate the plant at a point (x, y) somewhere within the region S, depicted in figure 1–1. The sole market for the paperboard is at city E, where the product can be sold at a price, λ_0. Let z_0 be the amount of product produced (and sold), and $c_0(x, y)$ be the cost per unit to transport the product from plant site (x, y) to city E.

There is considerable flexibility in terms of the fiber inputs that can be combined to produce the desired product; numerous grades of scrap paper, scrap paperboard, and wood chips can be mixed in a variety of ways to make the product. Wood chips are available only at location A (e.g., a lumber mill) at a price p_{1A}. Suppose only three grades of scrap paper/paperboard are usable; scrap 2 is available at location E, scraps 3 and 4 are available at both B and C; and scraps 2, 3, and 4 are all available at D. So there are four physically different fiber inputs (wood chips and three scrap papers), but because they are available from different sites at different prices they will be treated as nine distinct inputs. Let λ_{ij} be the price of fiber input i from source j, z_{ij} be the amount of fiber i from source j, and $c_{ij}(x, y)$ be the per unit cost of shipping fiber i from site j to plant location (x, y). Large amounts of power are also needed to make paperboard; for simplicity assume only electric power is of interest. Let S be divided into three regions, each of which is served by a different electric utility company. Within each region the cost of electricity is uniform, but across regions the costs differ. Let p_{5j} be the unit cost of electricity in region j, and z_{5j} be the amount of electricity used from region j. So there are three energy inputs, and the "transport" cost of energy is

$$c_{5j}(x, y) = \begin{cases} 0 & \text{if } (x, y) \text{ is in region } j \\ \infty & \text{if } (x, y) \text{ is not in region } j \end{cases}$$

Labor is available from sites B, C, and E at prices p_{6B}, p_{6C}, p_{6E}; a travel premium of $c_{6j}(x, y)$ must be paid to get labor from site j to work at location (x, y). Let z_{6j} be the amount of labor used from location j. Finally, let z_{7S} be the amount of capital equipment used, and suppose that capital equipment is available anywhere in S at price p_{7S}.

The available production technologies are described by a production function, F:[1]

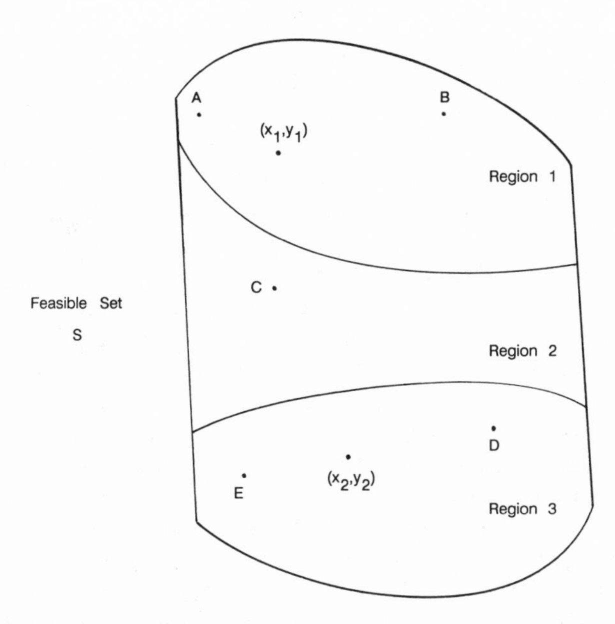

Figure 1–1. Location Space for Paperboard Plant Example

$$z_0 = F(z_{1A}, z_{2D}, \ldots, z_{4D}, z_{51}, \ldots, z_{53}, z_{6B}, \ldots, z_{6E}, z_{7S})$$

Although the specific form of F is not stated here, it should be recognized that many of the inputs are substitutable for each other (scrap 2 for wood chips, capital for labor, electricity for labor, etc.), and others are complementary (more wood chips may require more electricity, for example). These relationships are assumed to be captured by F.

The firm's problem is then to

$$\underset{z_{ij},(x,y)}{\text{maximize}} \qquad \pi = (p_0 - c_0(x,y))F(\mathbf{z}) - \sum_{i,j}(p_{ij} + c_{ij}(x,y))z_{ij} \qquad (1.1)$$

$$\text{subject to} \qquad z_{ij} \geqslant 0 \qquad \text{for all } (i,j) \qquad (1.2)$$

$$(x,y) \in S \qquad (1.3)$$

Notice that (1.1)–(1.3) integrates into one model the firm's problems of how much to produce, how to produce it, and where to produce it.

At each location (x, y) the firm faces different "delivered" prices for its inputs and output. Consequently, the optimal amount of paperboard to produce and the mix of inputs to use at, say location (x_1, y_1), will normally be different than at location (x_2, y_2). For example, because of its proximity to A it might be optimal at (x_1, y_1) to use larger amounts of wood chips and little or none of scrap 2, and labor would come primarily from cities B and C, whereas at location (x_2, y_2) it might be optimal to use large amounts of scrap 2 and little or no wood chips, and labor would probably come from city E. The amounts of labor, capital, and electricity would also vary from location to location according to their relative substitutability or complementarity and their prices.

Identifying the optimal production method and plant size a priori is not possible because it depends on the location of the plant. Likewise, the optimal location of the plant should not be determined a priori because it depends on the amount of inputs to be shipped and where they must be shipped. For example, a plant designed to use all three scraps in large quantities and no wood chips is probably best located near location D, whereas a plant designed to use primarily wood chips would probably be best near location A. Clearly, the production and location decisions of the firm are interrelated and need to be solved simultaneously. Separate optimization of the production problem—for an asumed set of delivered prices—and the location problem—for an assumed production level and technique—will usually lead to an *overall* decision that is suboptimal. In fact, Moses [1958, p. 272] warns that the a priori separation of these problems

> . . . tends to restrict, to an unwarranted extent, the range of possible types of locations which individual plants may consider. It may also delimit too severely the range of possible types of industries which underdeveloped areas are advised to consider in their development plans.

Relationship Between Production/Location and Other Location Models

Most of the well-known location models are special cases or simplifications of the integrated production/location model. In order to appreciate the generality and complexity of the integrated production/location model it is useful to compare it briefly with two other very common types of location models: Steiner-Weber models and generalized fixed-charge models. We begin the discussion, however, by first formulating a very general deterministic version of the production/location problem.

Suppose a firm produces one output from n spatially distinct inputs (the same physical input at different source locations is treated as different inputs), where the possible technologies are described by a production function, F. F is usually assumed to be quasiconcave and $F_i = \partial F/\partial z_i > 0$

for all i.[2] This latter assumption implies continuous substitutability of inputs. The assumpton of at least partial substitutability of inputs is fundamental to the work here; it is this substitutability that makes integration of the production and location problems meaningful. Although complete substitutability among all inputs is not necessary for the subsequent discussion to be valid (only partial substitutability of at least two inputs is needed), this assumption will be made for simplicity. The firm wishes to locate K new plants within some feasible set of locations, S, and will supply customers at M spatially distinct locations. Let x^k be the location of the kth plant, z_{0m}^k be the amount of output produced at plant k and sent to customers at market m, and z_i^k be the amount of input i used at plant k; so

$$\sum_{m=1}^{M} z_{0m}^k = F(z_1^k, \ldots z_n^k)$$

Let $c_{0m}(x^k)$ be the per unit cost of shipping the output from plant location x^k to market m, and $c_i(x^k)$ be the per unit cost of shipping input i to plant location x^k. Let p_{0m} be the selling price of the output at market m, and p_i be the price of input i at its source. The firm's production/location problem is then to

(PL)

$$\underset{z_{0m}^k, z_i^k, x^k}{\text{maximize}} \quad \pi = \sum_{k=1}^{K} \left[\sum_{m=1}^{M} (p_{0m} - c_{0m}(x^k)) z_{0m}^k - \sum_{i=1}^{n} (p_i + c_i(x^k)) z_i^k \right] \qquad (1.4)$$

$$\text{subject to} \quad \sum_{m=1}^{M} z_{0m}^k = F(z_1^k, \ldots, z_n^k) \qquad \text{for } k = 1, \ldots, K \qquad (1.5)$$

$$z_{0m}^k \geqslant 0,\ z_i^k \geqslant 0 \qquad \text{for all } m, i, k \qquad (1.6)$$

$$x^k \in S \qquad \text{for all } k \qquad (1.7)$$

Although we rarely consider such general problems in practice, this formulation should clarify the comparisons made in this chapter.

The Steiner-Weber Problem

As pointed out by Francis and White [1974, p. 186], the Steiner-Weber problem (also known as the general Fermat problem) has a long history. For the case of three, equally weighted, "demand" locations it was posed as a problem in pure geometry by Fermat in the early seventeenth century and solved by Toricelli prior to 1640. It was studied by the Swiss mathematician Steiner in the nineteenth century and by Weber, a German economist, early in the twentieth century. But it was not until the work of Weiszfeld [1937], Kuhn and Kuenne [1962], and Cooper [1963] that the problem could be considered essentially solved.

In its modern, plant-location form the problem can be stated as follows. Suppose a firm wishes to open a new plant which produces a product (output) from n inputs. The inputs are supplied from spatially distinct locations with coordinates (a_i, b_i), $i = 1, \ldots, n$ and the output is sent to a market with coordinates (a_0, b_0). There is a cost per unit shipped per unit distance for shipping each input, r_i, and the output, r_0. The rate of output to be produced is z_0 and of each input, i, is z_i. The rate of output is assumed to be given, and the technology is assumed to be represented by a Leontief (fixed intput/output ratio) production function, so that $z_i = s_i z_0$ for $i = 1, \ldots, n$, where the $s_i > 0$ are constants. Let (x, y) be the location of the firm's plant, and $d_i(x, y) = [(a_i - x)^2 + (b_i - y)^2]^{1/2}$ be the distance from the plant to (a_i, b_i), $i = 0, 1, \ldots, n$. The firm's problem is then to choose the values of (x, y) and the z_i's so as to minimize total cost:

$$\text{(PSW)} \quad \begin{aligned} &\underset{(x,y),z_i}{\text{minimize}} \quad C_{\text{PSW}} = \sum_{i=1}^{n} (r_i d_i(x, y) + p_i) s_i z_0 \\ &\qquad\qquad\qquad + r_0 d_0(x, y) z_0 \qquad (1.8) \\ &\text{subject to} \quad (x, y) \in R^2 \qquad (1.9) \end{aligned}$$

where R^2 is the real plane. Notice that because of the Leontief technology, with z_0 given, the quantities of each input used (z_i) are fixed as well and the terms $p_i s_i z_0$ in (1.8) are constants. Consequently (PSW) can be rewritten in its more common form:

$$\text{(SW)} \quad \begin{aligned} &\underset{(x,y)}{\text{minimize}} \quad C_{\text{SW}} = \sum_{i=0}^{n} r_i d_i(x, y) z_i \qquad (1.10) \\ &\text{subject to} \quad (x, y) \in R^2 \qquad (1.9) \end{aligned}$$

[(SW) can be generalized easily to include several plant locations and output markets and different distance metrics.]

When F is Leontief, transport costs are linear in Euclidean distance, and one facility is to be located our general production/location problem, (PL), reduces to (SW). The Leontief production function exhibits constant returns to scale and allows *no substitution of inputs*. Consequently, once the output level is given, there is no decision needed regarding the production variables. Thus (PL) reduces to the pure location problem (SW), and the firm's total costs are minimized (and profits maximized) by minimizing the transportation costs for the given input and output levels. It is this reduction that has lead many location theorists to conclude and suggest that the optimal location for a plant is the location that minimizes the firm's transport costs. Although true for (SW), this is *not* necessarily true if inputs are substitutable; in fact, if output level is not fixed, the minimum transport cost does not even make sense. However, the optimal location for (PL) *will* minimize transport costs *for the optimal values of the production variables* [Emerson 1973; Thisse and Perreur 1977]. So if the optimal values of the production variables in (PL) are known a priori, (PL) reduces to (SW) with the z_i's and z_0 fixed at their optimal values.

Although very restrictive in its production assumptions, (SW) is spatially general in the sense that the firm usually is assumed to be able to locate anywhere in the plane R^2, or a convex subset of R^2. (For normal metrics it can be shown that the optimal location will lie in the convex hull of $\{(a_0, b_0), (a_1, b_1), \ldots, (a_n, b_n)\}$; see Wendell and Hurter [1973a].) This convexity (and thus, continuity) of feasible locations allows the use of differential methods for solution. This fact forms the foundation of the established solution methods for versions of (SW), such as Weiszfeld [1937], Kuhn and Kuenne [1962], Cooper [1963, 1967], and Planchart and Hurter [1975]; these will be discussed in more detail in chapter 4.

Fixed-charge Type Problems

Operations researchers have been especially active in studying location problems that have been characterized as "fixed-charge" problems. The generalized fixed-charge problem incorporates the output level aspect of production into the facility location decision, but the input mix decisions either have been solved or fixed a priori. Unlike (SW), the set of feasible locations is usually a discrete set in space, and the problem generally is of interest only as a multifacility problem. Specifically, it is assumed that there are M spatially distinct customers, each demanding a quantity of output, D_m. There are J possible plant sites from which to choose; at each site the cost of establishing a plant and producing z_0^j units is given by a cost function, $C_j(z_0^j)$. Although the C_j's can have any form, they are usually assumed to have the following "fixed-charge" form:

$$C_j(z_0^j) = \begin{cases} 0 & \text{if } z_0^j = 0 \\ f_j + g_j z_0^j & \text{if } z_0^j > 0 \end{cases}$$

where f_j is a fixed cost of opening the plant at site j and g_j is a constant unit cost of production at j. Let S_j be the maximum production level possible at plant site j; if the S_j's are infinite then the problem is called an "uncapacitated fixed-charge" problem. Let r_{jm} be the cost to transport one unit of product from plant site j to customer m. (Note that there is no cost of shipping inputs to the plants explicitly included in the model; these are assumed to be captured in the cost functions.) The problem is then to select the number and locations of plants, the production level at each plant, and the shipping pattern so as to supply all customers with their desired demand, at minimum total cost. Thus, letting y_j be a variable that does "bookkeeping", where

$$y_j = \begin{cases} 0 & \text{if a plant is not opened at site } j \\ 1 & \text{if a plant is opened at site } j \end{cases}$$

for $j = 1, \ldots, J$, and letting z^j_{0m} be the number of units produced at site j and shipped to customer m, this problem can be stated mathematically as

$$\text{(FC)} \quad \begin{aligned} &\underset{y_j, z^j_{0m}}{\text{minimize}} && C_{\text{FC}} = \sum_{j=1}^{J} \left[f_j y_j + \sum_{m=1}^{M} (g_j + r_{jm}) z^j_{0m} \right] && (1.11) \\ &\text{subject to} && \sum_{j=1}^{J} z^j_{0m} = D_m \quad \text{for } m = 1, \ldots, M && (1.12) \\ & && \sum_{m=1}^{M} z^j_{0m} \leqslant S_j y_j \quad \text{for } j = 1, \ldots, J && (1.13) \\ & && z^j_{0m} \geqslant 0 && (1.14) \\ & && y_j \in \{0, 1\}. && (1.15) \end{aligned}$$

This problem is a 0–1 mixed-integer program. Numerous solution algorithms have been developed to solve various forms of (FC), including versions with general cost functions, C_j. Some of these will be discussed in chapter 3. (Problem (FC) is a special case of the class of spatial combinatorial problems known as location-allocation problems. An extensive literature on these problems exists; e.g., see Scott [1971], Ghosh and Rushton [1987], or Love, Morris and Wesolowsky [1988].)

Problem (FC) can be thought of as a discretized simplification of (PL). The fact that cost functions are available for each site implies that the input mix (production technique) decision, including the shipment of inputs, has been made or determined a priori for each feasible site. That is, the input mix decision for all possible output levels at all feasible sites is assumed to be known; so (PL) reduces to a problem of determining only the optimal location of each plant, the size (output level) of each plant, and the shipping plan for supplying customers. (FC) is more general than (SW) because input substitution is allowed (different input/output ratios can be represented in the C_j's), and output levels are variable. A theoretical limitation of (FC) is that it usually obscures the interaction of the production technology and plant location. Problem (FC), however, is computationally more tractable than (PL), and it will be shown later that under realistic assumptions (PL) sometimes can be reduced to a form of (FC).

In summary then, (PL) is the general problem faced by the firm when locating one or more new facilities. If the set of possible plant sites is finite, then a cost function (i.e., using optimal levels of each input) can be derived for each feasible site which includes all input transport costs. If this is done then (PL) reduces to a form of (FC); specifically, if the available technology generates cost functions that take the form of a fixed charge plus a constant times output, then (PL) reduces to (FC). If the input and output levels (or input/output ratios) in (PL) are fixed a priori (e.g., because F is Leontief in form) and if the location space is the plane, then (PL) reduces to a form of the pure location problem (SW).

Plan of the Book

Even though (PL) is often difficult, and sometimes impossible, to solve exactly, a priori separation of the production and location problems is not justified. Complete technical and spatial generality is not necessary for (PL) to provide valuable conclusions and insight. By looking at special cases of the production/location problem we are able to make the problem more tractable while obtaining an understanding of the interrelationships between the production and location variables. For some simpler special cases we are able to characterize the properties of an optimal solution and to identify the effects of changes in parameters—such as, prices, output, and transport costs—on the firm's optimal decision. In addition, the spatial and technological effects of governmental policies such as the rate and form of taxes, business incentives, and regulations are more easily identified for special cases of (PL). The results for these special cases suggest conjectures for more general situations, which can then be supported numerically, if not proven analytically.

The structure of this book is based upon the preceding philosophy. We begin with very simple models and then add greater generality. In chapter 2 the most fundamental forms of the production/location problem are presented. The set of possible plant locations is a line segment connecting the sources of inputs or the output market and an input source. Two of the principal solution properties, endpoint optimality and solution invariance, are first introduced for this simple case. The location space is generalized, in chapters 3 and 4, to networks and planes. Most of the solution properties developed for the linear space are shown to carry over into these more general spaces. The more general spaces make multiplant problems relevant. Multiplant models and solution properties are presented and compared with other location models.

In chapters 5 and 6 parametric uncertainty and risk preferences are introduced into the models. The models in these chapters identify the spatial and technological effects of parametric uncertainty. The mathematical complexity of the models is increased: however, some, but not all, of the solution properties that hold for the deterministic cases still hold under uncertainty, although additional assumptions are sometimes needed.

A serious drawback of production/location models is their mathematical complexity and the resulting difficulty in solving such problems. In chapters 3–6 solution procedures are presented and compared with methods used to solve (SW), (FC), and other location problems. Node optimality properties established in chapters 3 and 6 sometimes make it possible to reduce general production/location problems to variations of (FC). Likewise, a special separability property for the deterministic forms of (PL) is established in chapter 4, which makes the planar problem solvable using methods similar to those used for (SW).

Chapters 7 and 8 incorporate government policies, such as taxes, incentives, and regulations, into the firm's production/location decision. These chapters best illustrate the potential value of the production/location framework. In them it is shown that careful planning can often make government policies mutually beneficial to both the firm and the government. These two chapters may be the most important in the book, and so they were written to be relatively self-contained and less mathematical than the other chapters. In chapter 9 future directions for production/location research are suggested.

Notes

1. Although physically identical inputs from different locations are treated as different inputs in the argument of F, the actual technological description of the production pro-

cess (i.e., the functional form of F) would contain terms such as $(z_{1B} + z_{2C} + z_{2E})$ and $(z_{51} + z_{52} + z_{53})$, which combine physically identical inputs.

2. Quasiconcavity of F simply means that the production isoquants are convex to the origin; equivalently, the production-level sets are convex. In most cases F is strictly quasiconcave, meaning the isoquants are strictly convex.

2 DETERMINISTIC PRODUCTION/LOCATION MODELS ON A LINE

In this chapter the most elementary production/location models are presented. The basic features of these models are that the firm is assumed to use two inputs to produce one output, and the set of feasible plant locations is either assumed to be a line segment, or more generally, can be shown to be a line segment for the given assumptions regarding the locations of the input sources and output market. There are several reasons for beginning with this apparently simplistic case. First, most production/location research has dealt with this case. Second, it is a natural introduction to the more general, but complicated, network and planar problems. Many of the results that hold for more general models can be proved, illustrated, and explained more easily for the linear case, and then extended to the more complicated cases using technical mathematical properties. Third, the simpler structure of the location on a line problem makes it easier to study the interrelationships of the production and location decisions. Some theorems that can be obtained for two inputs and location on a line cannot be obtained for general networks or the plane. In particular, much of the parametric analysis cannot be demonstrated as strongly and precisely for the more general models. However, these results and observations for the linear case suggest reasonable conjectures for the more general versions which, although mathematically unproved (and possibly unprovable), can be investigated numerically. Finally, although simplistic, the two-input model on a line is of some interest in its own right. Quite often the firms production problem is dominated by two spatially distinct inputs, such as labor located in an urban area and a raw material available away from the city.

Models With No Output Transport Costs

Model Formulation and Example

Consider a firm that produces one output from two inputs where the production possibilities are described by a production function, F; that is, $z_0 = F(z_1, z_2)$ where z_0 is the rate of output and z_1 and z_2 are the usage rates of the inputs. It is assumed that F is quasiconcave and that $F_i \equiv \partial F/\partial z_i > 0$ for $i = 1, 2$. Let p_0, p_1, p_2 be the prices of the output and inputs at their markets and sources, respectively.[1] (For now all prices are assumed to be given.) In this section it is assumed that there is no cost for transporting output to market. Notice that this is equivalent to assuming that (a) customers are uniformly ubiquitous so the location of the plant does not affect output transport costs, or (b) all feasible locations are essentially the same distance from the market so that output transport costs are the same for all locations, or (c) customers will come to the plant to buy the product or will pay for all shipping costs regardless of plant location, or (d) the costs of transporting output are infinitesimal. Suppose the source of the first input is at location M_1 and the source of the second input is at location M_2. If s is the distance between M_1 and M_2, then the two points define a line segment and coordinate system $[0, s]$, where 0 is the location M_1 and s is the location M_2 (see figure 2–1). Let $c_1(x)$, $c_2(x)$ be the per unit costs of transporting the inputs from their source locations to the plant location, $x \in [0, s]$. (Notice that $c_1'(x) > 0$, $c_2'(x) < 0$.) It is not difficult to prove that if the transport costs are nonnegative and nondecreasing with distance, the firm's optimal plant location, x, will be on the segment $[0, s]$ [see Wendell and Hurter 1973a]. Then the firm's problem, which is a special case of (PL), is to choose the input levels z_1, z_2 (and, implicitly, z_0) and the plant location x, so as to maximize profit:

$$\text{(PL2.1)} \quad \begin{aligned} &\underset{z_1, z_2, x}{\text{maximize}} \quad && \pi = p_0 F(z_1, z_2) - (p_1 + c_1(x))z_1 \\ & && \qquad - (p_2 + c_2(x))z_2 && (2.1) \\ &\text{subject to} && z_1 \geqslant 0,\ z_2 \geqslant 0 && (2.2) \\ & && x \in [0, s] && (2.3) \end{aligned}$$

Often the output level is fixed in advance, i.e., $z_0 = \bar{z}_0$, in which case

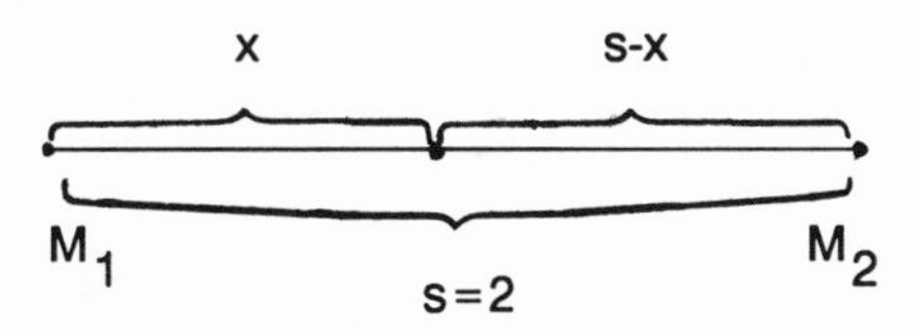

Figure 2–1. Location Space for Line Segment Models

the revenue part of equation (2.1) is fixed. In this case problem (PL2.1) can be written as the following cost minimization problem:

(DL2.1)

$$\underset{z_1, z_2, x}{\text{minimize}} \quad C = (p_1 + c_1(x))z_1 + (p_2 + c_2(x))z_2 \tag{2.4}$$

$$\text{subject to} \quad F(z_1, z_2) = \bar{z}_0 \tag{2.5}$$

$$z_1 \geqslant 0, \; z_2 \geqslant 0 \tag{2.2}$$

$$x \in [0, s] \tag{2.3}$$

Problems such as (DL2.1), in which output level is fixed, will be referred to as "design/location" problems because the firm must really choose only the input ratio and plant location; the absolute levels of the inputs are implied by the ratio z_1/z_2 if z_0 is given.

In this chapter several variations of (DL2.1) and (PL2.1) will be investigated. The properties of the optimal solutions will depend greatly upon the assumptions regarding the transport cost functions and the production function. To make the problem more tangible, consider the following numerical example of (DL2.1).

Example 2.1

Let $\bar{z}_0 = 4$, $p_1 = p_2 = 1$, $F(z_1, z_2) = (z_1 z_2)^{0.5}$, $s = 2$ (see figure 2–1), $c_1(x) = 0.5x$, and $c_2(x) = 0.5(2 - x)$; that is, the transport cost rates are liner in distance shipped. This firm's problem is then

$$\text{minimize} \quad C = (1 + 0.5x)z_1 + (1 + 0.5(2 - x))z_2$$

$$\text{subject to} \quad (z_1 z_2)^{0.5} = 4$$

$$z_1, z_2 \geqslant 0, 0 \quad 0 \leqslant x \leqslant 2$$

The optimal production solutions are listed below for three locations.

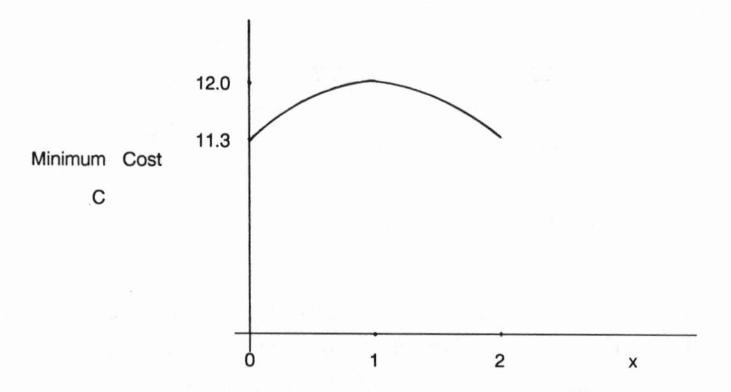

Figure 2–2. Minimum Cost as a Function of Plant Location for Example 2.1

	M_1	A	M_2
x	0	1	2
z_1	$2\sqrt{8}$	4	$\sqrt{8}$
z_2	$\sqrt{8}$	4	$2\sqrt{8}$
C	11.3	12	11.3

The two endpoints are better locations than the midpoint, A, for which $x = 1$. This can be seen more generally in figure 2–2 which plots minimum total cost as a function of x. Minimum total cost is a strictly concave function of x in this case, and the minimum of a concave function must occur at a boundary. In this specific case both boundary points are minima because the problem is symmetrical; both inputs have the same prices and transport cost functions, and the same "productivity" in F.

Suppose the problem were not symmetrical; e.g., let $p_1 = 2$ and $p_2 = 1$. The minimum total cost as a function of x is still strictly concave (see figure 2–3), but the location $x = 2 = M_2$ is now a unique optimum and $z_1^* = (4/3)\sqrt{3}$, $z_2^* = 4\sqrt{3}$, $C^* = 8\sqrt{3} = 13.9$.

Endpoint Optimality

The previous example leads to the following questions. Must (when must) an endpoint location be optimal? Can an interior point of the line be optimal? Why does endpoint optimality occur?

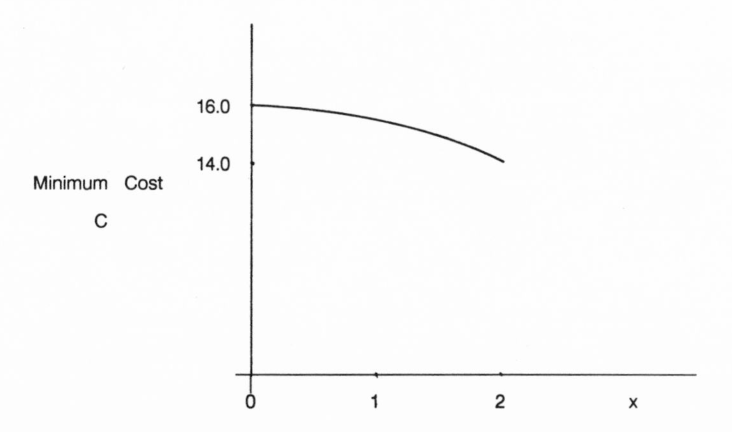

Figure 2–3. Minimum Cost as a Function of Plant Location for Modification of Example 2.1

Concave and Linear Transport Cost. Usually the cost of shipping a unit of a good increases with distance shipped, but the cost increases at a decreasing (though not necessarily strictly decreasing) rate. Furthermore, there is frequently some initial fixed cost which occurs once a good is transported at all (e.g., loading costs). Thus, c_1 and c_2 are often concave and discontinuous at $x = 0$ and $x = s$, respectively; figure 2–4 illustrates some typical transport cost functions.

Suppose that transport costs are concave in distance, but not linear over all of $[0, s]$; that is, for any x such that $0 \leqslant x = \lambda s \leqslant s$, where $0 \leqslant \lambda \leqslant 1$,

$$c_i(x) = c_i(\lambda s + (1 - \lambda)0) > \lambda c_i(s) + (1 - \lambda)c_i(0), \qquad i = 1, 2 \tag{2.6}$$

Now for *any* given input levels, including the optimal levels z_1^*, z_2^*, the objective functions of (PL2.1) and (DL2.1) are optimized by minimizing transport costs. So for *any* x, $0 < x < s$,

$$c_1(x)z_1^* + c_2(x)z_2^* > \lambda(c_1(s)z_1^* + c_2(s)z_2^*) \\ + (1 - \lambda)(c_1(0)z_1^* + c_2(0)z_2^*) \tag{2.7}$$

Now suppose

$$c_1(s)z_1^* + c_2(s)z_2^* = c_1(0)z_1^* + c_2(0)z_2^* \tag{2.8}$$

then the left-hand side (l.h.s.) of (2.7) > l.h.s. of (2.8) = r.h.s. of (2.8)—i.e., the transport cost of locating at either endpoint is less than locating at

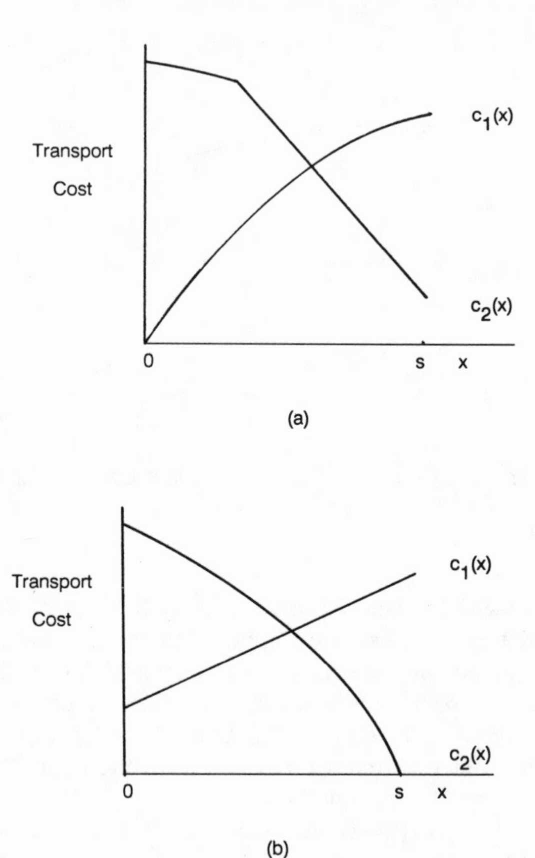

Figure 2–4. Typical Transport Cost Functions, $c_i(x)$

x. Likewise, if l.h.s. (2.8) $\lesseqgtr$ r.h.s. (2.8), then l.h.s. (2.7) $>$ l.h.s. (2.8) [r.h.s. (2.8)]; in either case, an endpoint location will be strictly better than x. Thus, we have proved the following:

Theorem 2.1

If the transport costs are concave in distance (with "jumps" allowed for c_1 at $x = 0$ and for c_2 at $x = s$), but not linear, then the optimal plant location for problems (DL2.1) and (PL2.1) can *only* be an endpoint of $[0, s]$. □

This result is the node optimality result which has been developed for pure location problems on networks that minimize total transport cost (network equivalents of (SW); see [Levy 1967; Wendell and Hurter 1973b]).

When transport costs are linear in distance, i.e., $c_1(x) = r_1 x$ and $c_2(x) = r_2(s - x)$, where $r_1, r_2 \geq 0$ are constants, Hakimi [1964] has shown that an endpoint location will always be optimal, but interior points could be optimal as well. More specifically, an interior location can be optimal if and only if every location on the line segment, including the endpoints, is optimal [Wendell and Hurter 1973b]. But when input substitutability exists, as in (PL2.1) and (DL2.1), a stronger result can be proved.

We begin by considering problem (DL2.1). The Lagrangian for (DL2.1) is

$$
\begin{aligned}
L = {} & (p_1 + c_1(x))z_1 + (p_2 + c_2(x))z_2 \\
& + \lambda(\bar{z}_0 - F(z_1, z_2)) + \mu(s - x) \qquad (2.9)
\end{aligned}
$$

The Kuhn-Tucker necessary conditions for an optimum are

$$
\left.
\begin{aligned}
\partial L/\partial z_i &= p_i + c_i(x^*) - \lambda^* F_i \geq 0 \qquad (2.10) \\
z_i^*(\partial L/\partial z_i) &= 0 \qquad (2.10a)
\end{aligned}
\right\} \quad i = 1, 2
$$

$$\partial L/\partial x = c_1'(x^*)z_1^* + c_2'(x^*)z_2^* - \mu^* \geq 0 \qquad (2.11)$$

$$x^*(\partial L/\partial x) = 0 \qquad (2.11a)$$

$$\partial L/\partial \lambda = \bar{z}_0 - F(z_1^*, z_2^*) = 0 \qquad (2.12)$$

$$\partial L/\partial \mu = s - x^* \geq 0 \qquad (2.13)$$

$$\mu^*(\partial L/\partial \mu) = 0 \qquad (2.13a)$$

$$\lambda^* \geq 0, \mu^* \leq 0 \qquad (2.14)$$

where an asterisk * denotes the optimal value of a variable.

The expression

$$c_1'(x^*)z_1^* + c_2'(x^*)z_2^* \qquad (2.15)$$

in (2.11) is the marginal total cost with respect to changes in location for the optimal input levels (this will be referred to as the "net locational pull" (NLP)). (A more extensive discussion of this marginal cost is given in chapter 4, where comparisons with the Weber problem are of greater interest.) From (2.11), (2.11a), (2.13), and (2.13a) it can be deduced (see Martinich and Hurter [1982] or Martinich [1980]) that

$$\text{if } (2.15) > 0, \qquad \text{then } x^* = 0 \qquad (2.16)$$

and

$$\text{if } (2.15) < 0, \qquad \text{then } x^* = s \tag{2.17}$$

therefore,

$$0 < x^* < s \quad \text{only if} \quad (2.15) = 0 \tag{2.18}$$

(for example, suppose (2.15) < 0. Then from (2.11), $\mu^* < 0$ and (2.13a) requires (2.13) = 0, so $x^* = s$.)

Theorem 2.2

If in problem (DL2.1) transport costs are linear in distance—i.e., ($c_1(x) = r_1x$ and $c_2(x) = r_2(s - x)$), then the optimal location can *only* occur at an endpoint of $[0, s]$.

Proof: (i) If either $z_1^* = 0$ or $z_2^* = 0$, the proof is trivial. If $z_1^* > 0$ and $z_2^* = 0$, then $x^* = 0$, and if $z_1^* = 0$ and $z_2^* > 0$, then $x^* = s$. (ii) So suppose $z_1^* > 0$, $z_2^* > 0$. From (2.16)–(2.18), it is sufficient to show that (2.15) $= r_1z_1^* - r_2z_2^* = 0$ is impossible. Assume $r_1z_1^* - r_2z_2^* = 0$. Then the minimum cost is given by

$$\begin{aligned} C^* &= (p_1z_1^* + r_1x^*z_1^* + p_2z_2^* + r_2sz_2^* - r_2x^*z_2^*) \\ &= p_1z_1^* + p_2z_2^* + r_2sz_2^* \end{aligned} \tag{2.19}$$

which is independent of the location variable x. Thus, for *any* fixed location $x \in [0, s]$, (z_1^*, z_2^*) must be an optimal solution to the subproblem:

$$\text{(P)} \qquad \begin{aligned} &\underset{z_1, z_2}{\text{minimize}} && (p_1 + r_1x)z_1 + (p_2 + r_2(s - x))z_2 \\ &\text{subject to} && F(z_1, z_2) = \bar{z}_0 \\ &&& z_1 \geqslant 0, z_2 \geqslant 0 \end{aligned}$$

Consequently, for every $x \in [0, s]$, z_1^* and z_2^* must satisfy the following necessary condition for (P):

$$\frac{(p_1 + r_1x)}{p_2 + r_2(s - x)} = \frac{F_1(z_1^*, z_2^*)}{F_2(z_1^*, z_2^*)} \tag{2.20}$$

The r.h.s. of (2.20) is constant in x, but the numerator of the l.h.s. is increasing in x and the denominator of the l.h.s. is decreasing in x. Thus, (2.20) can hold at most at one point and *not* for all $x \in [0, s]$. This is a contradiction, and so (2.15) = 0 must be impossible. □

The proof given for theorem 2.2 is not especially intuitive, but this approach is used here because it is easily extended to more complicated production/location problems on lines and networks. The essence of the proof, however, is that $r_1 z_1^* - r_2 z_2^*$ cannot equal zero for an optimal solution, because if it did, one of the inputs could be substituted for the other in a way that total cost could be reduced by locating at an endpoint. Specifically, suppose $r_1 z_1^* - r_2 z_2^* = 0$ and $x^* \in (0, s)$; then z_1^*, z_2^* must satisfy (2.20), i.e.,

$$\frac{F_1(z_1^*, z_2^*)}{F_2(z_1^*, z_2^*)} = \frac{p_1 + r_1 x^*}{p_2 + r_2(s - x^*)} \tag{2.21}$$

and the minimum total cost is

$$C^* = p_1 z_1^* + p_2 z_2^* + r_2 s z_2^* \tag{2.22}$$

(Note that $F_1(z_1^*, z_2^*)/F_2(z_1^*, z_2^*) = -(dz_2/dz_1)$ at (z_1^*, z_2^*).) Now there exists $\hat{z}_1 = z_1^* + \delta$, $\hat{z}_2 = z_2^* - \alpha\delta$, such that $F(\hat{z}_1, \hat{z}_2) = \bar{z}_0$ and

$$\frac{F_1(\hat{z}_1, \hat{z}_2)}{F_2(\hat{z}_1, \hat{z}_2)} = \frac{p_1}{p_2 + r_2 s} \tag{2.23}$$

where $\delta > 0$ and $\alpha = (z_2^* - \hat{z}_2)/\delta = (z_2^* - \hat{z}_2)/(\hat{z}_1 - z_1^*) = -(z_2^* - \hat{z}_2)/(z_1^* - \hat{z}_1)$. Therefore, α represents the negative of the slope of the line connecting the points (z_1^*, z_2^*) and $(\hat{z}_1, \hat{z}_2)$, and the l.h.s. of (2.21) and (2.23) represent the negative of the slopes of the isoquant at these two points, respectively. But because the isoquant is convex (because F is quasiconcave), it follows that (2.23) $< \alpha <$ (2.21).

Then the total cost of the solution $(\hat{z}_1, \hat{z}_2, x = 0)$ is

$$p_1 z_1^* + p_1\delta + p_2 z_2^* - p_2\alpha\delta + r_2 s z_2^* - r_2 s\alpha\delta \tag{2.24}$$

So (2.24) $<$ (2.22) if and only if

$$[p_1 - \alpha(p_2 + r_2 s)]\delta < 0 \tag{2.25}$$

But (2.23) $< \alpha$ implies that (2.25) holds, and so there is a solution at $x = 0$ that is better than the assumed interior optimum. If z_2 were increased instead, then the same argument can be made for $x = s$.

Two other approaches have been used to prove theorem 2.2. Shieh [1983b] has shown that the (minimum) cost function for *fixed* locations, $C(x, z_0)$, is a strictly concave function of x, and so its minimum can occur only at an endpoint of $[0, s]$, as in figures 2–2 and 2–3. (Sakashita [1967] and Emerson [1973] used the same approach but they assume F is homogeneous.) This concavity approach holds more generally. Using an

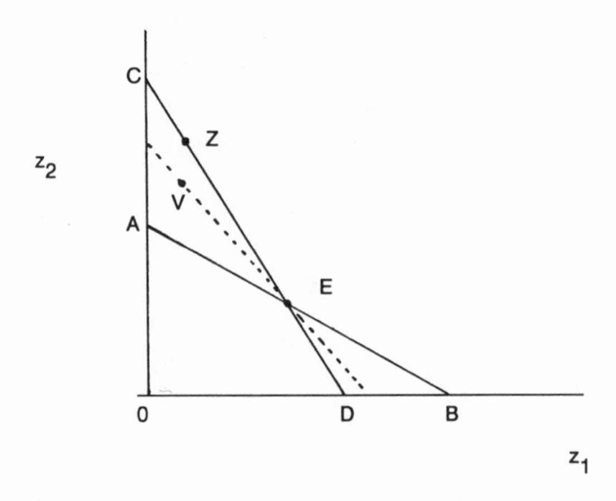

Figure 2–5. Iso-outlay Lines for Problem (DL2); *AB* Is Iso-outlay line at M_1 and *CD* IS Iso-outlay line at M_2

argument similar to that of Shephard [1970, ch. 4] it can be shown that if a firm's production function is strictly quasiconcave (all isoquants strictly convex), then the cost function $C(p; z_0)$ is strictly concave in the (delivered) input prices, p. Thus, C is strictly concave over any affine subset of the price space. When $x \in [0, s]$ the feasible delivered prices as a function of location, $p_1 + r_1x$ and $p_2 + r_2(s - x)$, implicitly define an affine subset of prices, and so $C(x, z_0) = C(p(x), z_0)$ is strictly concave in x. (The "dual approach" of Eswaran, Kanemoto, and Ryan [1981] is based upon this idea; they show how this approach can often lead to stronger results using more elegant methods.)

A graphical approach, which is simpler to see for location on a line but is less easily generalizable, has been used by Moses [1958], Emerson [1973], and Woodward [1973]. Each location on $[0, s]$ defines a ratio of *delivered* input prices: $(p_1 + r_1x)/(p_2 + r_2(s - x))$. This ratio defines the slope of the system of iso-outlay lines at that location. Two iso-outlay lines, representing the same total expenditure, say the minimum cost C^*, are shown in figure 2–5.

Line AB is an iso-outlay line associated with production at M_1, and CD is an iso-outlay line associated with production at M_2, representing the same total outlay as AB. Notice that more of input 1 can be obtained when the plant location is at M_1 $(x = 0)$ because the relative price of input 1 is less than when the plant is at M_2; the opposite is true for input 2. The iso-outlay line for *any* point $x \in (0, s)$, with total expenditure of C^*, will have its endpoints between AC and DB and will pass through point E. The boundary CEB represents the set (envelope) of efficient purchases for C^*.

(For example, the points V and Z represent the same expenditure at locations $x \in (0, s)$ and s, respectively. For the same expenditure, Z represents the same amount of input 1 as V but more of input 2.) If the isoquants of F are strictly convex (which will be true if F is quasiconcave and continuously differentiable), the tangency between the isoquant $z_0 = \bar{z}_0$ and the efficient envelope CEB cannot occur at E; thus, only an endpoint can be optimal (the isoquant could be tangent to CEB in two places, in which case both endpoints are optimal, as in figure 2–2).

When the output level is a variable, as in problem (PL2.1), theorems 2.1 and 2.2 still hold. (Although the shape of $p_0(z_0)$ *may* affect the optimal location, the strong endpoint-optimality property holds regardless of $p_0(z_0)$.) This follows immediately from the fact that for the optimal output level, z_0^* (a finite optimum is assumed to exist), (PL2.1) reduces to (DL2.1) with $\bar{z}_0 = z_0^*$. That is, for this output level the optimal input levels and plant location for (PL2.1) are those values that are optimal for (DL2.1) when $\bar{z}_0 = z_0^*$. The preceding can then be summarized as follows:

Theorem 2.3

For problems (DL2.1) and (PL2.1), if input transport costs are concave in distance (with "jumps" allowed at the input sources), then the optimal plant location can occur *only* at an endpoint of $[0, s]$. □

Nonconcave Transport Costs. When the transport costs c_1 and c_2 are not concave in distance the endpoint optimality of Theorem 2.3 no longer holds in general. Although endpoints may still be optimal even with nonconcave transport costs, the following example shows that interior points of $[0, s]$ can be unique optimal locations.

Example 2.2

Again consider example 2.1 except now let the transport cost functions be $c_1(x) = (x^2/4)$ and $c_2(x) = (2 - x)^2/4$. The minimum total cost as a function of x is given in figure 2–6, and the solutions for three locations on $[0, 2]$ are given below.

	M_1	A	M_2
x	0	1	2
z_1	$2\sqrt{8}$	4	$\sqrt{8}$
z_2	$\sqrt{8}$	4	$2\sqrt{8}$
C	11.3	10.0	11.3

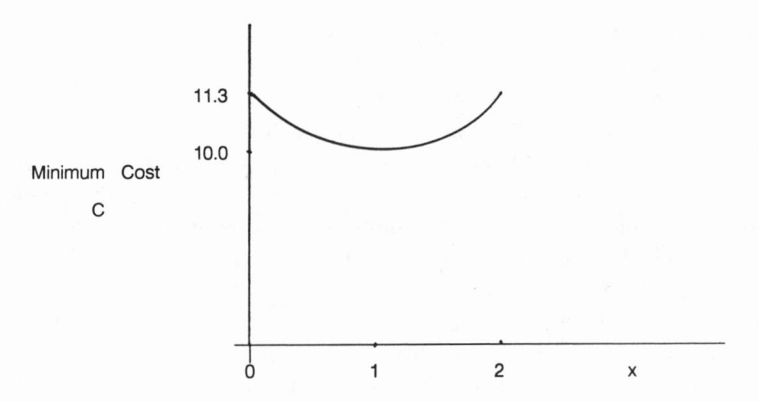

Figure 2–6. Minimum cost as a function of plant location for Example 2.2

In this case the optimal location is the interior point, $x^* = 1$, and the optimal values of the input variables are $z_1^* = z_2^* = 4$, $C^* = 10$.

As stated previously, the cost function $C(p; z_0)$ is concave in the delivered prices, p. But the delivered prices are functions of x. So when transport costs are concave in distance (including linear), the cost function as a function of x, $C(x; z_0) = C(p(x); z_0)$, is (strictly) concave in x because the concave composition of a concave function is concave. But when transport costs are convex (or more generally nonconcave) in distance then it is possible that the composition of the concave function, $C(p)$, and the nonconcave function, $p(x)$, is nonconcave, as shown in the preceding example. When $C(x; z_0)$ is not concave in x, then it is possible that the "net locational pull," $c_1'(x^*)z_1^* + c_2'(x^*)z_2^*$, could equal zero so that the economic forces of transportation of the two inputs can exactly balance at an interior location of $[0, s]$. When $C(x)$ is strictly concave such a balance is impossible and so an interior optimum is impossible.

Models With Output Transport Costs

Model Formulations

Output transport costs have been introduced into the two-input production/location model on a line in two ways. The first approach [e.g., Sakashita 1967] assumes that one of the inputs is ubiquitous, so that its delivered cost is independent of plant location. The optimal plant location is then restricted to the line segment connecting the source of the nonubiquitous (transportable) input and the market for the output. An alternative model (e.g., see Mathur [1979]; Higano [1985]) assumes that both inputs are transportable with the source of input 1 being at location M_1 and the source of input 2 and the output market being at location M_2, a distance s from M_1. This defines a coordinate system $[0, s]$ with $M_1 = 0$ and $M_2 = s$. Let $c_i(x)$ be the per unit cost to transport input i from its source to plant location $x \in [0, s]$, and $c_0(x)$ be the cost per unit to transport the output from x to its market (note that $c_0'(x) < 0$). The first model, with a ubiquitous input, is really a special case of the latter model; specifically, if $c_2(x) = 0$ for all x then input 2 can be treated as a ubiquitous input, and the second model reduces to the first model. Therefore, we will restrict our attention to models of the second type.

Let $p_0(z_0)$ be the price of the output at its market, as a function of z_0 (if price is independent of the firm's output level, as in a competitive market, then $p_0(z_0) = p_0$). The firm's problem is then

(PL2.2)

$$\text{maximize} \quad \pi = [p_0(F(\mathbf{z})) - c_0(x)]F(\mathbf{z}) - (p_1 + c_1(x))z_1 - (p_2 + c_2(x))z_2 \tag{2.26}$$

$$\text{subject to} \quad z_1 \geqslant 0,\ z_2 \geqslant 0 \tag{2.2}$$

$$x \in [0, s] \tag{2.3}$$

If output is fixed a priori so that $z_0 = F(z_1, z_2) = \bar{z}_0$, then $p_0(z_0) = p_0$, and (PL2.2) becomes the cost minimization design/location problem

(DL2.2)

$$\text{minimize} \quad C = c_0(x)\bar{z}_0 + (p_1 + c_1(x))z_1 + (p_2 + c_2(x))z_2 \tag{2.27}$$

$$\text{subject to} \quad z_1 \geqslant 0,\ z_2 \geqslant 0 \tag{2.2}$$

$$x \in [0, s] \tag{2.3}$$

$$F(z_1, z_2) = \bar{z}_0 \tag{2.5}$$

Endpoint Optimality

If all the transport cost functions in (PL2.2) and (DL2.2) are concave but not linear over all of $[0, s]$, then the argument used to prove theorem 2.1 can be applied to show that *only* endpoints of $[0, s]$ can be optimal plant locations for these problems; this also follows from Wendell and Hurter [1973b]. Similarly, examples can be constructed which show that if at least one of the transport cost functions is not concave in x, then it is *possible* for an interior location to be optimal. The remaining issue then is whether only endpoints can be optimal if the transport costs are linear in x, i.e., when $c_0(x) = r_0(s - x)$, $c_1(x) = r_1 x$, $c_2(x) = r_2(s - x)$.

We begin by considering problem (DL2.2), and then follow the same approach as that used to prove theorem 2.2. If we construct the Lagrangian for (DL2.2), the necessary conditions for an optimum are

$$\partial L/\partial z_1 = p_1 + r_1 x^* - \lambda^* F_1(z_1^*, z_2^*) \geqslant 0 \tag{2.28}$$

$$\partial L/\partial z_2 = p_1 + r_2(s - x^*) - \lambda^* F_2(z_1^*, z_2^*) \geqslant 0 \tag{2.29}$$

$$\partial L/\partial x = -r_0 \bar{z}_0 + r_1 z_1^* - r_2 z_2^* - \mu^* \geqslant 0 \tag{2.30}$$

In this case the net locational pull at the optimal location is

$$-r_0 \bar{z}_0 + r_1 z_1^* - r_2 z_2^* \tag{2.31}$$

It can be shown that if (2.31) > 0 then x^* must equal 0, and if (2.31) < 0 then x^* must equal s; so $x^* \in (0, s)$ only if (2.31) = 0.

Theorem 2.4

Suppose the transport cost functions in (DL2.2) are linear in x. If $z_2^* > 0$, then the optimal plant location can only be at an endpoint of $[0, s]$.

Proof: If $z_1^* = 0$, then clearly $x^* = s$. So let $z_1^* > 0$. From the comments above it is sufficient to show that (2.31) cannot equal 0. We will prove by contradiction; so assume that (2.31) = 0. Then the minimum cost is given by

$$C^* = p_1 z_1^* + p_2 z_2^* + (r_0 \bar{z}_0 + r_2 z_2^*)s \tag{2.32}$$

which is independent of the location variable, x. The remainder of the proof is then identical to that for theorem 2.2. □

It is now straightforward to prove the following theorem.

Theorem 2.5

Suppose the transport cost functions in (PL2.2) are linear in x. If $z_2^* > 0$, then the optimal plant location can only be at an endpoint of $[0, s]$.

Proof: Let z_0^* be the optimal output level (we assume a finite optimum exists). Then the firm's optimal input levels and location must be optimal for (DL2.2) with $\bar{z}_0 = z_0^*$. The result then follows immediately from theorem 2.4. □

The requirement in theorems 2.4 and 2.5 that z_2^* be greater than zero suggests two questions: (1) How restrictive is that requirement, and (2) Why is this requirement necessary? The answer to the first question is that it is not very restrictive. Most commonly used two-input production functions, such as the Cobb-Douglas production function, have the property that $F(z_1, 0) = F(0, z_2) = 0$. Consequently, if z_0 is positive then z_2^* must be positive, and x^* will be an endpoint. The second question is best answered by looking at an example.

Example 2.3

Consider problem (DL2.2) where $p_1 = p_2 = 2$, $r_0 = 1$, $r_1 = 4$, $r_2 = 4$, $\bar{z}_0 = 4$, $F(z_1, z_2) = 4z_1 + 2z_2$, and $s = 1$. The optimal solution is then $z_1^* = 1$, $z_2^* = 0$, and x can be any location on $[0, s]$; the minimum cost is 6.

So if $z_2^* = 0$, then it is *possible*, although very unlikely, for an interior location to be optimal because (2.31) will equal zero if and only if $z_1^* = (r_0/r_1)\bar{z}_0$. Even in this case, however, the endpoints will be optimal as well, and so a search for an optimum can be restricted to only the endpoints.

Parametric Analysis

Production/location models are especially useful in determining how changes in the environment (i.e., changes in the parameters) will affect the firm's production and location decisions. Three types of parametric changes will be studied using the models developed earlier in this chapter: changes in output level, changes in prices, and changes in transport cost rates.

Changes in Output; Spatial and Production Invariance

In practice, when firms are deciding where to locate a plant and which production method to use, they often perform their analysis assuming some desired output capacity; that is, they solve a design/location problem—such as (DL2.1) or (DL2.2). Of course the best output capacity is usually not known with certainty a priori, and so it is common to solve the "design/location" problem for several possible output levels, z_0. In the design/location context, a reasonable question then is whether a perturbation in the fixed output levels, $\bar{z}_0$, will cause a change in the optimal input ratio and plant location, and if so, how? The following example demonstrates that a change in $\bar{z}_0$ *can* alter the optimal input mix and location for (DL2.1).

Example 2.4

Let $p_1 = p_2 = 2$, let $s = 2$, let $c_1(x) = x$, $c_2(x) = 2 - x$, and $F(z_1, z_2) = z_1(z_2 + 2) - (z_1^2/4)$, where $z_2 > (z_1/2) - 2$. The optimal solutions for (DL2.1) for $\bar{z}_0 = 2, 4, 8, 16, 24$ are given below.

$\bar{z}_0$	z_1^*	z_2^*	z_1^*/z_2^*	x^*
2	1.172	0.000	∞	0
4	2.309	0.309	7.48	0
8	3.266	1.266	2.58	0
16	4.619	2.619	1.76	0
24	3.266	6.165	0.53	2

The production function in example 2.4 has the property that input 2 is not needed to produce small amounts of output. However, in order to produce large amounts of output, input 2 is needed in increasingly greater proportions. Although production functions of this type can occur in practice, the more common types of production functions have the property that they are homothetic or can be approximated by a homothetic function. A production function, $F(z_1, \ldots, z_n)$, is *homothetic* if and only if it can be written in the form $F(z_1, \ldots, z_n) = F(\mathbf{z}) = f(g(\mathbf{z}))$ where $g(\mathbf{z})$ is a homogeneous production function of degree one and f is a transform function—i.e., f is a finite, nonnegative, upper-semicontinuous, nondecreasing function with $f(0) = 0$. (For an extensive discussion of homothetic functions, see Shephard [1970].) Homogeneous production functions, such as the Cobb-Douglas and Arrow-Chenery-Minhas-Solow production functions, are a subset of the family of homothetic functions. But other, nonhomogeneous, production functions such as

$$F(\mathbf{z}) = \begin{cases} a + \sum_{i=1}^{n} b_i \log z_i & \text{if } a + \sum_{i=1}^{n} b_i \log z_i \geq 0 \\ 0 & \text{otherwise} \end{cases}$$

are homothetic as well.

The distinctive property of homothetic production functions is that their isoquants are "parallel" in the following sense. Let z_0^1 and z_0^2 be two positive levels of output. Then if F is homothetic, there exists some scalar $v > 0$ such that for all input vectors $\mathbf{z}^1$ with $F(\mathbf{z}^1) = z_0^1$, $F(v\mathbf{z}^1) = z_0^2$; likewise, for every $\mathbf{z}^2$ with $F(\mathbf{z}^2) = z_0^2$ it follows that $F((1/v)\mathbf{z}^2) = z_0^1$. (A well-known consequence of this property in nonspatial economics is that the firm's expansion path is a straight line when input prices are constant.) This property can be used to prove the following theorem [Hurter, Martinich, and Venta 1980, theorem 1].

Theorem 2.6

In (DL2.1) if F is homothetic, then the optimal plant location x^* and the optimal input ratio, z_1^*/z_2^*, are independent of the output level, $\bar{z}_0$, and so any location that is optimal for one output level is optimal for all output levels.

Proof: Define $T(z_0) = \{(\mathbf{z}, x) \mid F(\mathbf{z}) = z_0$ and $(\mathbf{z}, x)$ is optimal for (DL2.1) with output level $z_0\}$. Let z_0^1 and z_0^2 be two arbitrary output levels. Let $(\mathbf{z}^1, x^1)$ be an optimal solution for $z_0 = z_0^1$, and $(\mathbf{z}^2, x^2)$ be optimal for $z_0 = z_0^2$. Then for some scalar v, such that $F((1/v)z_0^2) = z_0^1$,

$$\sum_{i=1}^{2} (p_i + c_i(x^1))z_i^1 \leqslant \sum_{i=1}^{2} (p_i + c_i(x^2))(z_i^2/v) \tag{2.33}$$

or

$$\sum_{i=1}^{2} (p_i + c_i(x^1))vz_i^1 \leqslant \sum_{i=1}^{2} (p_i + c_i(x^2))z_i^2 \tag{2.34}$$

Thus, $(v\mathbf{z}^1, x^1) \in T(z_0^2)$—i.e., $(v\mathbf{z}^1, x^1)$ must be an optimal solution for (DL2.1) with $z_0 = z_0^2$. It can be shown similarly that $((1/v)\mathbf{z}^2, x^2) \in T(z_0^1)$. It follows that $T(z_0^2) = \{(v\mathbf{z}^1, x^1) \mid (\mathbf{z}^1, x^1) \in T(z_0^1)\}$, and the desired result follows immediately. □

The properties in theorem 2.6 are referred to as "spatial (or location) invariance" and "production invariance." It is evident from the proof of theorem 2.6 that these results hold for more general design/location problems; specifically, they hold when there are n inputs and they apply to *any* location space, not just a line. (See Khalili, Mathur, and Bodenhorn [1974]; Hurter, Martinich, and Venta [1980]; and Eswaran, Kanemoto, and Ryan [1981]. For brevity, these references will sometimes be referred to as KMB [1974], HMV [1980], and EKR [1981], respectively.)

To illustrate theorem 2.6, again consider example 2.2. The optimal solution for $\bar{z}_0 = 4$ is $x^* = 1$, $z_1^* = 4$, $z_2^* = 4$. The optimal solution for any output level $v\bar{z}_0$ is $x^* = 1$, $z_1^* = z_2^* = 4v$. Notice that this invariance property does not require that the optimal location be an endpoint, and it holds for any form of transport cost functions, $c_i(x)$ (see Woodward [1973] and Mathur [1979]).

Eswaran, Kanemoto, and Ryan [1981] and Hurter and Venta [1982] have provided an alternative proof to the locational invariance property in theorem 2.6 which emphasizes a form of separability resulting from the homotheticity of F. They have shown that if F is homothetic the cost function $C(x, z_0)$ (i.e., the minimum total cost, including transportation, of producing z_0 units of output at location x) can be written as

$$C(x, z_0) = h(x)g(z_0) \tag{2.35}$$

Thus, the location that minimizes total cost for one output level must minimize total cost for all output levels. Although the "separability" of (2.35) makes it possible to identify the spatial invariance, the real value of (2.35) appears to be in solving more general production/location problems, and so this separability will be addressed in detail later.

If the planned (optimal) output level is not known a priori, theorem 2.6 identifies a very common condition under which problem (DL2.1) can be solved for *any* fixed output level to determine the optimal plant location and the optimal input ratio. If output is, in fact, a variable, as in (PL2.1), then the cost function $C(x^*, z_0)$ can be derived directly and the optimal output level determined separately.

When output transport costs are included in the problem, as in (DL2.2), location and production invariance can still occur, but stronger production function restrictions are required.

Theorem 2.7

If F is homogeneous of degree one, then the optimal plant location and input mix for (DL2.2) are invariant with output level. Specifically, if $(\mathbf{z}^1, x^1)$ is optimal for output level z_0^1, then $((z_0^2/z_0^1)\mathbf{z}^1, x^1)$ is optimal for any output level z_0^2.

Proof: Let $(\mathbf{z}^1, x^1)$ be optimal for output level z_0^1, let $(\mathbf{z}^2, x^2)$ be optimal for output level z_0^2, and define $t = z_0^2/z_0^1$. By definition of degree one homogeneity, $F(t\mathbf{z}^1) = z_0^2$ and $F(z^2/t) = z_0^1$. Now by the optimality of $(\mathbf{z}^1, x^1)$

$$C^*(z_0^1) = c_0(x^1)z_0^1 + \sum_{i=1}^{2}(p_i + c_i(x^1))z_i^1$$
$$\leq c_0(x^2)z_0^2/t + \sum_{i=1}^{2}(p_i + c_i(x^2))(z_i^2/t) \qquad (2.36)$$

or

$$c_0(x^1)tz_0^1 + \sum_{i=1}^{2}(p_i + c_i(x^1))(tz_i^1) \leq \sum_{i=1}^{2} c_0(x^2)z_0^2 + (p_i + c_i(x^2))z_i^2$$
$$= C^*(z_0^2) \qquad (2.37)$$

So $(t\mathbf{z}^1, x^1)$ is optimal for (DL2.2) with $z_0 = z_0^2$. □

(For other approaches to theorem 2.7 see Emerson [1973], Woodward [1973], KMB (1974), Mathur [1979], HMV [1980], and EKR [1981].)

If F is not homogeneous of degree one then it is possible for the optimal solution of (DL2.2) to change with $\bar{z}_0$. Even when F is homogeneous, the optimal location and input ratio can change if the degree of homogeneity is not one. This is shown in the following example.

Example 2.5

In (DL2.2) let $p_1 = p_2 = 2$, $s = 2$, $c_0(x) = 1.5(2 - x)$, $c_1(x) = x$, $c_2(x) = (2 - x)$, and $F(z_1, z_2) = z_1^{0.6}z_2^{0.2}$. The optimal solution for $\bar{z}_0 = 4$, 8, and 12 are given below.

$\bar{z}_0$	z_1^*	z_2^*	x^*
4	6.260	4.173	2
8	14.888	9.926	2
12	34.955	5.826	0

In example 2.5, at any *fixed* location the optimal input ratio is the same regardless of $\bar{z}_0$, because F is homothetic. For example, letting $z_i(x)$ be the optimal value of z_i at location x, then at $x = 0$, $z_1(0)/z_2(0) = 6$; at $x = 2$, $z_1(2)/z_2(2) = 1.5$. The net locational pull at x, normalized with respect to $\bar{z}_0$, is

$$[-1.5\bar{z}_0 + z_1(x) - z_2(x)]/\bar{z}_0 = (z_1(x)/\bar{z}_0) - (1.5 + z_2(x))/\bar{z}_0) \qquad (2.38)$$

But because F exhibits decreasing returns to scale, as $\bar{z}_0$ increases, the input/output ratios, $z_i(x)/\bar{z}_0$ increase (in the same proportion), so that the input transport costs become relatively more important and the output transport costs become relatively less important. The "pull" toward $x = 0$ increases relative to that toward $x = s$ as $\bar{z}_0$ increases. For small values of $\bar{z}_0$, the pull toward the output market dominates and $x^* = s$; but as $\bar{z}_0$ increases, the relatively greater use of input 1 becomes sufficiently strong so that at large values of $\bar{z}_0$ the optimal location shifts to $x^* = 0$.

A number of researchers (e.g., KMB [1974], Mathur [1979]) have shown that this phenomenon holds more generally.

Theorem 2.8

In problem (DL2.2), suppose F is homogeneous of degree k. Let $(\mathbf{z}^j, x^j)$ be optimal for output level z_0^j, and let $z_0^2 > z_0^1$. Then (i) x^2 $(\geqslant, =, \leqslant)$ x^1 and (ii) z_1^2/z_2^2 $(\leqslant, =, \geqslant)$ z_1^1/z_2^1 according as k $(>, =, <)$ 1.[2]

Proof (i): Let $z_i^h(x)$ be defined as the optimal value of z_i when location is fixed at x and output level is fixed at z_0^h (so $z_i^h = z_i^h(x^h)$). Define

$$P^h(x^j) \equiv [p_1 + c_1(x^j)]z_1^h(x^j) + [p_2 + c_2(x^j)]z_2^h(x^j) \tag{2.39}$$

So $P^h(x^j)$ is the cost of purchasing and transporting the optimal amounts of inputs when the plant location is fixed at x^j and the output level is fixed at z_0^h. Now by the optimality of $(\mathbf{z}^1, x^1)$ for $z_0 = z_0^1$

$$P^1(x^1) + c_0(x^1)z_0^1 \leqslant P^1(x^2) + c_0(x^2)z_0^1 \tag{2.40}$$

or

$$A \equiv [c_0(x^1) - c_0(x^2)]z_0^1 \leqslant [P^1(x^2) - P^1(x^1)] \equiv B \tag{2.41}$$

Define $t = z_0^2/z_0^1 > 1$. Notice that $z_i^2(x) = t^{1/k}z_i^1(x)$ for any fixed x and any k, so $P^2(x^j) = t^{1/k}P^1(x^j)$. Then by the optimality of $(\mathbf{z}^2, x^2)$ for $z_0 = z_0^2$,

$$t^{1/k}P^1(x^2) + c_0(x^2)tz_0^1 \leqslant t^{1/k}P^1(x^1) + c_0(x^1)tz_0^1 \tag{2.42}$$

or, rearranging terms,

$$A \geqslant t^{[(1/k)-1]}B \tag{2.43}$$

If $k < 1$, then $t^{[(1/k)-1]} > 1$, and (2.41) and (2.43) can both hold only if $A, B \leqslant 0$. But $A \leqslant 0$ if and only if $c_0(x^1) \leqslant c_0(x^2)$; that is $x^2 \leqslant x^1$. Similarly, if $k > 1$, then $t^{[(1/k)-1]} < 1$ and (2.41) and (2.43) can both hold only if $A, B \geqslant 0$. But $A \geqslant 0$ if and only if $x^1 \leqslant x^2$. The result for $k = 1$ is immediate from theorem 2.7, or can be shown to follow immediately from the fact that $t^{[(1/k)-1]} = 1$.

Proof (ii): The input-ratio results follow immediately from the location inequalities in (i), the homogeneity of F, and the fact that $z_1^h(x)$ is decreasing in x and $z_2^h(x)$ is increasing in x (these latter claims are proved formally in the following subsection). □

The preceding result has intuitive appeal. For example, as $\bar{z}_0$ increases, if $k < 1$, the amount of output per unit of input decreases. Thus, for any location, as the output level increases the economic pull of output transport cost decreases relative to the input transport costs, and so the pull toward the market will decrease relative to the pull toward $x = 0$. When $k > 1$ the opposite occurs. Mathur [1979, p. 306] implies that theorem 2.8 holds in general only if the output transport cost structure has special properties, but as shown here, it holds for any nonnegative output transport cost structure as long as the transport *rate* is independent of quantity.

The conditions of theorem 2.8 can, in fact, be relaxed for localized changes in output level. If F is homothetic then (i) and (ii) hold according as F exhibits locally (increasing, constant, decreasing) returns to scale; see EKR [1981] or Martinich and Hurter [1988].

If F is not homogeneous (or at least homothetic), then the effect of changes in output level cannot be determined in general. The optimal location could be affected both by changes in the input ratio (if F is not homothetic) and by fluctuating returns to scale. The overall effect of these conditions cannot be determined in general, but rather would have to be calculated on a case-by-case basis.

Changes in Prices

We begin our discussion by considering the effects of price perturbations on design/location problems. The qualitative effects of price changes on (DL2.1) and (DL2.2) are the same, so we will only consider the more general problem, (DL2.2). It is well known from microeconomics that for a two-input model and *fixed* location, x, $dz_i(x)/dp_i < 0$ and $dz_j(x)/dp_i > 0$ for $j \neq i$, where $z_i(x)$ is the optimal value of z_i at location x. The question of interest is how will $(\mathbf{z}, x)$ be affected by changes in prices?

Theorem 2.9

Consider either (DL2.1) or (DL2.2).

(a) Let $(\mathbf{z}^h, x^h)$ be optimal for prices (p_0, p_1^h, p_2). If $p_1^2 > p_1^1$, then (i) $z_1^2 \leqslant z_1^1$; (ii) $z_2^2 \geqslant z_2^1$; and (iii) $x^2 \geqslant x^1$.

(b) Let $(\mathbf{z}^h, x^h)$ be optimal for prices (p_0, p_1, p_2^h). If $p_2^2 > p_2^1$, then (i) $z_1^2 \geqslant z_1^1$; (ii) $z_2^2 \leqslant z_2^1$; and (iii) $x^2 \leqslant x^1$.

Proof (a): (i) Define P^{hk} to be the cost of purchasing and transporting inputs under prices (p_0, p_1^h, p_2) if the optimal solution under prices (p_0, p_1^k, p_2) is used; that is,

$$P^{hk} = (p_1^h + c_1(x^k))z_1^k + (p_2 + c_2(x^k))z_2^k \tag{2.44}$$

Then by the optimality of $(\mathbf{z}^1, x^1)$

$$P^{11} + c_0(x^1)z_0 \leqslant P^{12} + c_0(x^2)z_0$$

or

$$A \equiv [c_0(x^1) - c_0(x^2)]z_0 \leqslant P^{12} - P^{11} \equiv B \tag{2.45}$$

Similarly, by the optimality of $(\mathbf{z}^2, x^2)$,

$$P^{22} + c_0(x^2)z_0 \leqslant P^{21} + c_0(x^1)z_0$$

or

$$[c_0(x^1) - c_0(x^2)]z_0 \geqslant P^{22} - P^{21} \tag{2.46}$$

But $P^{22} = P^{12} + \alpha z_1^2$ and $P^{21} = P^{11} + \alpha z_1^1$, where $\alpha = p_1^2 - p_1^1$. So (2.46) can be written as

$$A \geqslant B + \alpha(z_1^2 - z_1^1) \tag{2.47}$$

But (2.45) and (2.47) can both hold only if $z_1^2 \leqslant z_1^1$.

(ii) $z_2^2 \geqslant z_2^1$ follows immediately from (i) because z_0 is fixed and all the $F_i > 0$.

(iii) If $z_i^1 = z_i^2$, then the result is immediate; so assume $z_i^1 \neq z_i^2$. Because x^1 is the optimal location for prices (p_0, p_1^1, p_2)

$$c_0(x^1)z_0 + c_1(x^1)z_1^1 + c_2(x^1)z_2^1 \leqslant c_0(x^2)z_0 + c_1(x^2)z_1^1 + c_2(x^2)z_2^1$$

or rearranging terms

$$[c_1(x^1) - c_1(x^2)]z_1^1 \leqslant [c_2(x^2) - c_2(x^1)]z_2^1 + [c_0(x^2) - c_0(x^1)]z_0 \tag{2.48}$$

Now defining $C =$ the term in the first [ ], $D =$ the term in the second [ ], and $E =$ the term in the third [ ], (2.48) can be written as

$$Cz_1^1 \leqslant Dz_2^1 + Ez_0 \tag{2.49}$$

Likewise, because x^2 is optimal for prices (p_0, p_1^2, p_2)

$$Cz_1^2 \geqslant Dz_2^2 + Ez_0 \tag{2.50}$$

Now if $x^2 < x^1$, then C, D, $E > 0$; so from (i) $Cz_1^1 > Cz_1^2$ and from (ii)

$Dz_2^2 > Dz_2^1$. But then (2.49) and (2.50) cannot both hold simultaneously. Thus $x^2 \geqslant x^1$. (Notice that if $x^2 \geqslant x^1$, then A, B, $C \leqslant 0$, and (2.49) and (2.50) are not inconsistent.)

Proof (b): The proof is identical to that of (a). □

It is tedious, but not difficult, to show that the inequalities in theorem 2.9 for the z_i^h's hold strictly if $z_1^1 > 0$ and $z_2^1 > 0$. It should also be noted that theorem 2.9 is independent of the transport cost structure (assuming that the costs increase with distance shipped). The qualitative effects of input price perturbations described by theorem 2.9 are exactly what one would expect intuitively. As the price of an input increases, use of that input decreases and use of the other input increases. The locational "forces" will then tend to push the optimal plant location away from the source of the input whch has had a price increase and toward the source of the other input. (This force, however, may not be sufficient to actually cause a change in the optimal location.)

Changes in the output price, p_0, obviously will not affect the optimal solution for the design/location problems because output is fixed (unless the price drops so low that the firm chooses to produce nothing). However, output price changes can affect the optimal solutions to the production/ location problems, and so we will begin our consideration of (PL2.1) and (PL2.2) by looking at the effects of output price perturbations.

If p_0 increases, for example due to an increase in market demand, the optimal output level will increase (if a smaller output level were optimal it would have been optimal originally).[3] The qualitative effects on the optimal values of z_1, z_2, and x are then the same as those caused by a parametric change in the output levels in (DL2.1) and (DL2.2). If F is not homothetic then it is impossible to determine, in general, the effects of an output price change on the optimal location and input levels. However, in practice F is often approximated by a homothetic function and sometimes by a homogeneous function, and so the following results are of interest. (It is assumed in the following that a finite optimum for each problem exists.)

Theorem 2.10

In (PL2.1) let $(z_0^h, z_1^h, z_2^h, x^h)$ be optimal for prices (p_0^h, p_1, p_2). Suppose F is homothetic and let $p_0^2 > p_0^1$. Then (i) $z_0^2 > z_0^1$; (ii) $z_1^2/z_2^2 = z_1^1/z_2^1$; and (iii) $x^2 = x^1$.

Proof: (i) is obvious. (ii) and (iii) follow immediately from (i) and theorem 2.6. □

Theorem 2.11

In (PL2.2) let $(\mathbf{z}^h, x^h)$ be optimal for prices (p_0^h, p_1, p_2), and let $p_0^2 > p_0^1$. If F is homogeneous of degree $k < 1$, then (i) $z_0^2 > z_0^1$, (ii) $z_1^2/z_2^2 \geq z_1^1/z_2^1$, and (iii) $x^2 \leq x^1$.

Proof: (i) is obvious. (ii) and (iii) follow immediately from theorem 2.8. □

(Theorem 2.11 is stated only in terms of $k < 1$ because if $k \geq 1$ then either the optimum is indeterminate or no finite optimum with $z_0 > 0$ exists.)

Theorems 2.8 and 2.11 identify the effects of output level and price perturbations on the optimal input *ratio*; however, notice that they do not make a statement about the optimal levels of input usage. The reason for this is that the inclusion of output transport costs in the model can lead to a very interesting phenomenon; spatially induced factor "inferiority." An input (factor) is defined to be *inferior* if its "use declines when output (and thus cost) is expanded at constant factor prices" [Ferguson and Saving 1969, p. 776]. As one might expect, factor inferiority is quite rare in practice; in fact, in a nonspatial setting it cannot occur with a homogeneous production function [Puu 1971, p. 243].[4] However, in spatial models such as (DL2.2) and (PL2.2) it is quite possible for the use of an input to decrease as the output level increases *even for a homogeneous production function*, as demonstrated in example 2.5. When output increases, as long as the optimal location does not change, the optimal input levels will all increase in the same proportion so that the ratio z_1^*/z_2^* remains constant. But if the optimal location changes, such as when z_0 goes from 8 to 12 in example 2.5, then the absolute use of input 2 can decrease. The reason for this is clear. The change in location causes a change in the *delivered* factor prices, so even through the factor prices at their sources do not change, the effective factor prices do change and they cause this apparent factor inferiority. One implication of this result is that when production functions are derived empirically using data from spatially dispersed plants, researchers need to be careful about using delivered factor prices in their analysis rather than "market" prices. Ignoring transportation costs can lead to anomalies, due to spatial factors, in the empirically derived production function which then will not accurately represent the true technological relationships.

The effects of input price perturbations on (PL2.1) and (PL2.2) can be derived easily from the earlier results. Unless the input is inferior (in a nonspatial sense), an increase in the price of the input will cause a decrease in the optimal output level because the marginal cost curve will shift upward [Samuelson 1976, p. 67]. If F is not homothetic then the effects on the input variables and location are ambiguous in general, so we will again look at the common cases, where F is homothetic or homogeneous.

Theorem 2.12

(a) In (PL2.1) let $(\mathbf{z}^h, x^h)$ be optimal for prices (p_0, p_1^h, p_2), and let $p_1^2 > p_1^1$. If F is homothetic, then (i) $z_0^2 < z_0^1$, (ii) $z_1^2/z_2^2 \leqslant z_1^1/z_2^1$, and (iii) $x^2 \geqslant x^1$. (b) For changes in p_2, (a) holds with the signs of (ii) and (iii) reversed.[5]

Proof (a): (i) is true by the assumed noninferiority of the inputs. Now by the homotheticity of F and theorem 2.6 there exists a scalar v such that $(v\mathbf{z}^1, x^1)$ is optimal for (DL2.1) with $z_0 = z_0^2$ and prices (p_0, p_1^1, p_2). Then by theorem 2.9, $z_1^2 \leqslant vz_1^1$, $z_2^2 \geqslant vz_2^1$ (so $z_1^2/z_2^2 \leqslant z_1^1/z_2^1$), and $x^2 \leqslant x^1$. Thus, (ii) and (iii) are proven.

Proof (b): Proof is identical to (a) with obvious sign changes. □

When output transport costs are included, somewhat weaker results occur.

Theorem 2.13

(a) In (PL2.2) let $(\mathbf{z}^h, x^h)$ be optimal for prices (p_0, p_1^h, p_2), and let $p_1^2 > p_1^1$. If F is homogeneous of degree $k < 1$, then (i) $z_0^2 < z_0^1$; (ii) $z_1^2/z_2^2 \leqslant z_1^1/z_2^1$; and (iii) $x^2 \geqslant x^1$. (b) For changes in p_2, part (a)(i) holds, but (ii) and (iii) do not—the effects on the input ratio and location are ambiguous.

Proof (a): The proof is essentially the same as for theorem 2.12a except that theorem 2.8 is used in place of theorem 2.6.

Proof (b): (i) is obvious. The ambiguity of the input ratio and location effects is due to two conflicting forces: (1) The increase in p_2 causes a direct substitution effect of input 1 substituting for input 2, thus increasing the "pull" toward $x = 0$. (2) The increase in p_2, however, causes output to decrease, which decreases the overall input/output ratio (because $k < 1$). Thus, the relative "pull" of the output market (i.e., toward $x = s$) increases; this creates a substitution of input 2 for input 1, as specified in theorem 2.8. The overall effect is ambiguous, in general. □

This section has shown that, for the most part, the effect of price changes on the optimal input ratio, output level, and plant location are exactly what one would expect. However, the inclusion of output transport costs can cause some very interesting phenomena to occur, such as the apparent factor inferiority of a technologically noninferior input.

Changes in Transport Rates

Two characteristics of transport costs have been shown to be of importance: the shape of the transport cost rate function and the actual magnitude of the transport costs. The first characteristic was shown to be crucial in determining whether interior locations of the line segment could be eliminated from consideration a priori. The sensitivity analyses of the preceding two subsections, however, are independent of the shape of the transport rate functions. Because *changes* in shape are difficult to characterize and because such perturbations are not especially common in practice, perturbations in shape are not considered here (to the best of our knowledge, changes in the shape of transport cost functions have not been studied in any location research).

The (relative) magnitudes of the transport rates are certainly of great importance in determining the optimal plant location. Increases and decreases in transport rates among the inputs and output can cause a change in the optimal input mix, output level, and location. On the surface it may appear that an increase in the per unit transport rate of an input would be equivalent to an increase in the source price of that input and, qualitatively, have the same spatial and production effects. In fact, the opposite is often the case. This is because unlike a change in the source price which affects the NLP only through its effect on the z_i^* values, changes in the transport rates can affect NLP directly through the $c_i'(x)$ terms as well. This can be seen most clearly by looking at a numerical example.

Example 2.6

Consider problem (DL2.1) with $z_0 = 150$, $s = 2$, $p_1 = p_2 = 20$, $c_1(x) = r_1 x$, $c_2(x) = (2 - x)$, $F(z_1, z_2) = z_1^{0.3} z_2^{0.6}$. We give the optimal solutions for two values of r_1 below.

r_1	x^*	z_1^*	z_2^*	C^*
2	2	145.94	350.27	10508.02
3	0	175.62	319.30	10536.94

When r_1 is only 2 it is more economical to locate at the source of the second input and transport the first input. However, as r_1 is increased to 3 (or more) the (marginal) cost of transporting the first input becomes large enough so that it pays to locate at its source and to ship the second input. Notice that the amount of input 1 used also increases; this again is an anomaly which would tend not to occur in nonspatial economics—as input 1 becomes relatively "more expensive" we actually increase its use. (This is because by changing plant locations we can actually reduce its delivered price relative to input 2.) Thus, even though *at any fixed location* the delivered price of input 1 increases, because the firm can control partially the delivered price it pays by adjusting its location, it is often advantageous to move toward the source of the increasingly expensive input and to use more of it.

Therefore, we cannot conclude that the optimal amount of an input will decrease if its transport rate increases; in fact, it can change in either direction. However, when input transport rates are linear in distance we can conclude the following.

Theorem 2.14

In (DL2.1) and (DL2.2) let $c_i(x) = r_i d_i$ where $r_i > 0$ and d_i is the distance from the source (market) for i to x. Let (z_1^h, z_2^h, x^h) be optimal for transport rates r_0, r_1^h, and r_2. Then if $r_1^2 > r_1^1$, $x^2 z_1^2 \leqslant x^1 z_1^1$. The same holds for changes in r_2 (except with $s - x$ replacing x).

Proof: The proof is essentially the same as for theorem 2.9, except that $\alpha = (r_1^2 - r_1^1)$; the details are omitted for brevity. □

Thus, it is possible for an increase in r_i to cause either an increase or decrease in z_i^*) however, an increase in r_i will cause a decrease in the total "quantity-miles" of input i, transported, $d_i z_i^*$ [Heaps 1982].

For changes in the output transport rate, stronger results are possible. These are achievable because the firm can reduce the effect of increases in the output transport rate only by moving closer to the output market; unlike for inputs, substitution options do not exist.

Theorem 2.15

Assuming the same conditions as in theorem 2.14, if $r_0^2 > r_0^1$, then (i) $x^2 \geqslant x^1$, and (ii) $z_1^2/z_2^2 \leqslant z_1^1/z_2^1$.

Proof: Proof is similar to theorem 2.9 and is omitted. □

The only important aspect of transport costs that has not been discussed here is the assumed linearity of transport costs in quantity shipped. That is, in all of the models considered here the costs of transportation have been of the form $c_i(x, z_i) = c_i(x)z_i$, where $c_i(x, z_i)$ is the cost of transporting z_i units of input i (output) from its source to location x (from location x to its market); there are no economies or diseconomies of scale in quantity shipped. The proofs of the endpoint optimality results of the preceding two sections did not rely on the linearity of transport costs in quantity shipped, and so those results hold without change even if transport economies or diseconomies exist (i.e., concavity in *distance* is the crucial characteristic). The invariance and other sensitivity results, however, do not necessarily hold if there are transport economies or diseconomies of scale, as the following example demonstrates.

Example 2.7

Consider problem (DL2.1) with $p_1 = p_2 = 1$, $F(z_1, z_2) = (z_1z_2)^{0.5}$, and $s = 2$. Let the transport costs $c_i(x)z_i$ be replaced by $c_i(x, z_i)$, where $c_1(x, z_1) = 0.5xz_1$ and $c_2(x, z_2) = (2 - x)z_2^{0.5}$. The optimal solutions for $\bar{z}_0 = 2$, 4, 8, and 16 are given below.

$\bar{z}_0$	z_1^*	z_2^*	x*	C*
2	1.414	2.828	2	5.657
4	2.828	5.657	2	11.314
8	9.410	6.802	0	21.428
16	17.996	14.225	0	39.764

For small output levels, and small values of z_2, input 2 is relatively expensive to transport, and so the optimal location is at its source, $x = 2$. As output level and the use of input 2 increase, the per unit shipping cost of input 2 decreases to the point where the optimal location shifts to $x = 0$. Notice also that even when the optimal location remains the same ($\bar{z}_0 = 8, 16$), the optimal input ratio can change because the relative delivered prices of the inputs change.

The complication caused by transport economies and diseconomies is that the spatial economic forces created by them can counteract the direct spatial effects underlying the sensitivity results. Even so, in many cases the spatial forces created by transport economies and diseconomies can reinforce the direct effects so that the theorems in the preceding subsections still hold. There are many specific cases when this can occur; however, any listing of cases is likely to be incomplete, and providing formal proofs for each is tedious. Instead, we will simply identify and explain two cases where the earlier sensitivity results still hold, and the reader can then use similar reasoning for specific cases encountered in practice.

Suppose the input transport costs are linear in quantity shipped, but the output transport costs are concave in quantity shipped, i.e., economies of scale. Then if $k \leqslant 1$, theorems 2.8 and 2.11 hold because, as output increases, its cost per unit shipped for any distance decreases, which reduces the locational pull toward the market for output at $x = s$, reinforcing the original spatial forces in those theorems. Similarly, if the transport costs for input 1 and the output are linear in quantity and the transport costs for input 2 are concave in quantity, then theorems 2.8, 2.9, and 2.11 hold for the same reason. In essence, if all transport costs are linear in quantity except one, then if that remaining transport cost function is either strictly concave or convex, the effect of quantity changes can usually be identified as being identical to price perturbations, which makes the analysis simpler.

Solution Methods

When transport cost functions are concave in distance shipped, theorems 2.1–2.5 make solution of problems (PL2.1), (PL2.2), and (DL2.2) relatively simple; the optimal location can only be at either $x = 0$ or $x = s$. Consequently, to solve these problems one can simply solve the appropriate production subproblem with x fixed equal to zero and then with x fixed equal to s; whichever solution yields the better objective function value is optimal. Even though the optimal location sometimes can be determined after solving only one subproblem, in general the subproblems at both endpoints must be solved to determine the complete optimum. For example, consider (DL2.1). If we let $z_i(x)$ be the optimal value for z_i when the plant location is fixed at x, then under normal conditions $dz_1(x)/dx < 0$ and $dz_2(x)/dx > 0$, so the net locational pull at x,

$$c_1'(x)z_1(x) + c_2'(x)z_2(x) \tag{2.51}$$

is nonincreasing in x because $z_1(x)$ is decreasing and $z_2(x)$ is increasing in x. Consequently, if (2.51) > 0 for $x = s$, then the optimal plant location must be $x^* = 0$, but the production subproblem at $x = 0$ must still be solved to determine the optimal values for z_1 and z_2. Similarly, if (2.51) < 0 for $x = 0$, then it is immediately known that $x^* = s$; however, the subproblem for $x = s$ would still have to be solved to obtain z_1^* and z_2^*. Notice that if $c_1'(0)z_1(0) + c_2'(0)z_2(0) > 0$ one cannot correctly conclude that $x^* = 0$ because it is possible that $c_1'(s)z_1(s) + c_2'(s)z_2(s) < 0$; likewise, knowing that (2.51) < 0 for $x = s$ provides no help in determining the optimal location without solving the subproblem at $x = 0$.

When transport cost functions are not all concave in distance, solution of the previous problems becomes far more difficult. (However, even though interior optimal locations are possible, endpoint optima are still very likely, so it is reasonable to begin any search for an optimum by solving the production subproblems at $x = 0$ and $x = s$.) One simplification which is theoretically easy but often difficult to use in practice is to prove that the minimum total cost function, $C(x)$, is concave in x. Even if some of the transport cost functions are nonconcave, $C(x)$ can be concave, in which case an endpoint must be optimal and so only the two endpoints need to be checked. In most cases, however, proving the concavity of $C(x)$ is very dificult and not worth the effort. Similarly, if $C(x)$ can be proved to be convex—or if $\pi(x)$ in (PL2.1) or (PL2.2) can be proved to be concave—the solution process can be simplified. But proving such properties is also usually not worth the effort.

In practice, the most efficient way to find a globally optimal location is to use a "grid search" method similar to that suggested by Nijkamp and Paelink [1973]. The line $[0, s]$ is divided into equal segments, $[0, s_1]$, $[s_1, s_2], \ldots, [s_m, s]$. At each endpoint of a segment the corresponding production subproblem is solved at that location, and the optimal cost or profit at that location is obtained. In addition, the first- and second-order conditions for an optimum are checked at each of these locations. From this information one of the segments is chosen and divided into a finer grid of segments, and the production subproblems are then solved. This is repeated until further dividing of segments does not produce significant improvement in the solution. Nijkamp and Paelink [1973] illustrate this method in the plane and so we will omit a numerical example for now.

In summary, computational solution of problems on a line segment is relatively easy. Solution methodology becomes a dominant issue, however, in succeeding chapters when additional complexities are added to the problem.

Summary and Conclusion

The models introduced in this chapter are the most elementary of the production/location models, yet they permit the development of many of the important features and properties of production/location theory. It was shown that when transport costs are concave in distance, interior points of the line segment can be excluded from consideration as optimal locations (these have been referred to as "exclusion theorems" [Higano 1985]). The approach used to prove endpoint optimality is adapted in the next chapter to establish "node optimality" properties for more general production/location problems on networks.

Throughout this chapter, whether or not output transport costs are included in the model significantly affects the properties of the optimum. When output transport costs are included, interior optima become a relatively rare, but possible, occurrence when transport costs are linear. In addition, it was shown that the optima for models with output transport costs are more easily altered by parametric perturbations than when output transport costs are zero.

Possibly the most important results of this chapter, are the sensitivity results. These results most clearly illustrate how the substitutability of inputs can affect the optimal location, and that each feasible location can have a different optimal production solution. The changes in the optimum resulting from changes in output level, prices, and transport costs are essentially what one would expect (if the person is familiar with production/location theory). They demonstrate that even small changes in parameters can often cause dramatic changes in the firm's optimal plant location, input mix, and output level. The most striking example of this is the apparent, "spatially induced," factor inferiority which can result from price or output changes, even when technologically the input is not inferior.

Much of the presentation in this chapter forms the foundation for subsequent chapters; however, the models presented here are important in their own right for two reasons. First, the simple nature of these models make it possible to achieve more precise and understandable results than are obtainable for more complex models. This provides the intuition and understanding regarding the interrelationships among the production and location variables necessary to make sense out of more complex and realistic situations. Second, in some cases the two-input model on a line is, in fact, realistic; many production processes are dominated by two "transportable" inputs, such as labor and a raw material, or two raw materials. Processes that include several inputs, of which only two vary in price over space, can sometimes be analyzed using the models in this chapter.

Notes

1. See note 2 in chapter 1 for a definition of quasiconcave.

2. This theorem is not stated or proved in terms of dx^*/dz_0, in order to make it more general; it applies to cases where the optimal locations are endpoints, as well as interior

locations, and F need not be everywhere differentiable. However, if the z_i^j's are strictly positive and F is differentiable then the input ratio inequality holds strictly.

3. The shape of $p_0(z_0)$ does not affect the subsequent results, but we do assume that $p_0'(z_0) \leqslant 0$.

4. Puu [1971] provides a description of when and how factor inferiority might occur in practice.

5. If the $z_i^h > 0$ then the input ratio inequality holds strictly.

3 DETERMINISTIC PRODUCTION/LOCATION MODELS ON NETWORKS

By restricting the production process to two inputs and the location space to a line segment, in chapter 2 we were able to investigate many of the salient features of the production/location interrelationships. Now the problem is generalized both in terms of the production variables and, more importantly, the feasible location set. In a practical sense most plant location decisions must be made with respect to some transportation network. Consequently, in this chapter we assume that the firm must make its production/location decision so as to locate on some given transport network. Any finite number of spatially distinct input and output markets are allowed, with their locations being at nodes of the network. First, a single-plant network problem is formulated along with a numerical example. Using methods similar to those in chapter 2, we then show that in many cases the set of optimal plant locations can be reduced to the node set of the network.

For network location spaces, multiplant problems become relevant, and so they are introduced next. It is shown that all of the node optimality properties that hold for single-plant problems also hold for comparable multiplant problems. These node-optimality properties make it possible to reduce the network production/location and design/location problems to variations of the fixed-charge and K-median problems (the latter is a multifacility, network version of the Weber problem).[1] This reduction allows the application of fixed-charge and K-median algorithms (after some preliminary computations), which greatly simplifies solution of the problems. This reduction and the relationships among these problems is included in the discussion on solution methods.

As noted earlier some of the parametric results developed in chapter 2 extend easily to network spaces (specifically, the invariance results for output perturbations); however, unambiguous specification of the effects of changes in prices or transport rates is quite difficult for network spaces. Those parametric results that can be established are presented in the section on parametric analysis, along with some conjectural observations illustrated through numerical examples.

Single-Plant Models

We will first formulate two production/location problems of increasing complexity (along with their design/location counterparts). In the following subsection we will then identify conditions under which the search for an optimal location can be restricted to the nodes of the network.

Formulation and Example

Consider a firm that produces one output from n spatially distinct inputs. (Two inputs which are physically identical except for source location are considered to be different inputs.) In the first model we will assume that output transport costs are nill or fixed so they can be ignored (e.g., customers will come to the plant to buy and location will not affect their purchase plans). The production relationship is again given by a quasiconcave production function, F, so $z_0 = F(z_1, \ldots, z_n)$ where z_0 is the rate of output and z_i is the usage rate of input i, and $F_i \equiv \partial F/\partial z_i > 0$ for all i. Let the input source locations be at nodes of a transport network $\eta = (N, A)$ where N is the set of nodes and A is the set of arcs. (If the input sources are not at nodes of the actual transport network, a new network is easily defined by adding these points to the node set and redefining the arcs.) For clarity, let the input source nodes be labelled $N_1, \ldots, N_n$; the other nodes of the network may correspond to warehouse locations, traffice intersections, etc. Let p_i be the price of input i at its source location and $p_0(z_0)$ be the price the firm receives for its output as a function of output level ($p_0'(z_0) \leq 0$).

For any arbitrary arc, A_{ab}, which connects nodes N_a and N_b, let $\delta_{ab} > 0$ be the length of the arc (for simplicity assume $\delta_{ab} = \delta_{ba}$). A distance scale is induced on the arc A_{ab} with $x = 0$ corresponding to N_a and $x = \delta_{ab}$ corresponding to N_b. Let ${}_{ab}\phi_i(x)$ be the per unit cost of transporting input i

from N_a to $x \in A_{ab}$. Let $c_i(A_{ab}, x)$ be the *lowest* cost of shipping one unit of input i from its source location, N_i, to $x \in A_{ab}$; notice that either[2]

$$c_i(A_{ab}, x) = c_i(A_{ab}, 0) + {}_{ab}\phi_i(x)$$

or

$$c_i(A_{ab}, x) = c_i(A_{ab}, \delta_{ab}) + {}_{ba}\phi_i(\delta_{ab} - x)$$

The firm's problem can then be written as

$$\text{(PL3.1)} \quad \begin{aligned} &\underset{z_i, x}{\text{maximize}} && \pi = [p_0(F(z_1, \ldots, z_n))]F(z_1, \ldots, z_n) - \sum_{i=1}^{n} [p_i + c_i(A_{ab}, x)]z_i \\ &\text{subject to} && z_i \geqslant 0 \qquad (3.1) \\ & && x \in \eta \qquad (3.2) \end{aligned}$$

where in choosing x, A_{ab} is implied to be chosen as well. If the output level is fixed at $z_0 = \bar{z}_0$ a priori, then (PL3.1) becomes the design/location problem

$$\text{(DL3.1)} \quad \begin{aligned} &\underset{z_i, x}{\text{minimize}} && C = \sum_{i=1}^{n} [p_i + c_i(A_{ab}, x)]z_i \\ &\text{subject to} && \bar{z}_0 = F(z_1, \ldots, z_n) \\ & && z_i \geqslant 0 \qquad (3.1) \\ & && x \in \eta \qquad (3.2) \end{aligned}$$

A generalization of these problems is to assume the firm is to supply several output markets, say M of them, each with its own inverse demand (price) function $p_{0m}(z_{0m})$, where z_{0m} is the amount of output sent to market m. We will also assume that the cost of shipping output is significant and borne by the firm. Let ${}_{ab}\phi_0(x)$ be the cost per unit to ship output from location $x \in A_{ab}$ to N_a and $c_{0m}(A_{ab}, x)$ be the lowest cost of shipping one unit of output from $x \in A_{ab}$ to market m—notice that $c_{0m}(A_{ab}, x)$ equals either $c_{0m}(A_{ab}, 0) + {}_{ab}\phi_0(x)$ or $c_{0m}(A_{ab}, \delta_{ab}) + {}_{ba}\phi_0(\delta_{ab} - x)$. In this case the firm's problem can be written as

$$\text{(PL3.2)} \quad \underset{z_i, z_{0m}, x}{\text{maximize}} \quad \pi = \sum_{m=1}^{M} [p_{0m}(z_{0m}) - c_{0m}(A_{ab}, x)]z_{0m} - \sum_{i=1}^{n} [p_i + c_i(A_{ab}, x)]z_i$$

$$\text{subject to} \quad \sum_{m=1}^{M} z_{0m} = F(z_1, \ldots, z_n)$$

$$z_i \geqslant 0 \tag{3.1}$$

$$x \in \eta \tag{3.2}$$

where the selection of x is implied to involve a selection of A_{ab} as well. Just as with (PL3.1), in some cases the quantities z_{0m} are set a priori equal to some levels $\bar{z}_{0m}$. In which case (PL3.2) becomes the design/location problem

$$\text{(DL3.2)} \quad \begin{aligned} \underset{z_i, x}{\text{minimizing}} \quad & C = \sum_{m=1}^{M} [c_{0m}(A_{ab}, x)]\bar{z}_{0m} + \sum_{i=1}^{n} [p_i + c_i(A_{ab}, x)]z_i \\ \text{subject to} \quad & \sum_{m=1}^{M} \bar{z}_{0m} = F(z_1, \ldots, z_n) \end{aligned}$$

$$z_i \geqslant 0 \tag{3.1}$$

$$x \in \eta \tag{3.2}$$

The purpose for explicitly defining four separate problems is that the characteristics of the optima for each are somewhat different, and we feel that it is worthwhile to identify the differences clearly (this will be done in the next two subsections). On the other hand, these four cases should capture the essence of most *single*-facility, network production/location problems.

To briefly illustrate a network production/location problem, we present the following numerical example of (PL3.1).

Example 3.1

Suppose a firm produces one output from two physically distinct inputs, A and B. Input A is available from nodes 1 and 3 at prices $p_1 = 3.5$ and $p_3 = 3.7$, and input B is available from nodes 2 and 4 at prices $p_2 = 4.4$ and $p_4 = 4.5$ (see figure 3–1). Let the price received by the firm for its output be independent of output level with $p_0 = 12$ and let $F(z_1, \ldots, z_4) = (z_1 + z_3)^{0.4}(z_2 + z_4)^{0.4}$. Output transport costs are assumed to be nill and the input transport costs for each input are 0.1 per unit per unit distance,

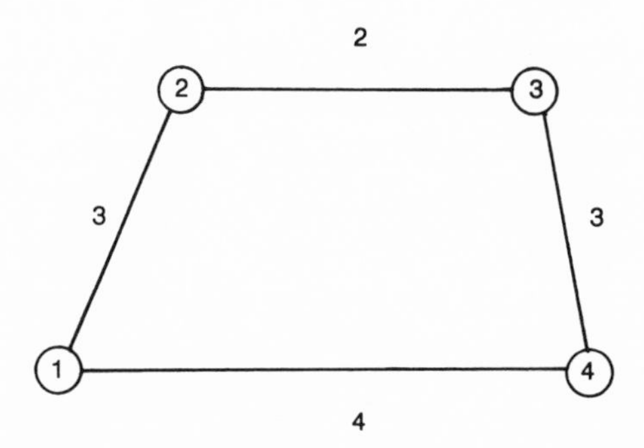

Figure 3–1. Network Space for Example 3.1

i.e., ${}_{ab}\phi_i(x) = 0.1x$ for all i and A_{ab}. The arc lengths are given in figure 3–1. The firm's optimal solution for this problem is to locate at node 1 with $z_0^* = 1.961$, $z_1^* = 2.690$, $z_2^* = 2.003$, $z_3^* = z_4^* = 0$, and $\pi^* = 4.703$.

Node Optimality

In example 3.1 the fact that the optimal location was at a node of the network was not a coincidence. It can be shown that under the most common circumstances an optimal plant location for (PL3.1)–(DL3.2) will be at a node and, under some circumstances, can *only* be at a node. Specifically, suppose the input and output transport costs along every arc are concave in distance shipped; that is, all ${}_{ab}\phi_i$, $i = 1, 1, \ldots, n$, are concave. (The ${}_{ab}\phi_i$ can be discontinuous as long as they are lower semi-continuous with "jumps" occurring only at nodes.) Figures 3–2(a) and 3–2(b) illustrate examples of these functions; the jumps at the nodes (such as in figure 3–2(b)) could represent a fixed loading or transfer charge. It follows that every $c_i(A_{ab}, x)$ and $c_{0m}(A_{ab,} x)$ is concave in x *on every arc*. Figure 3–3 illustrates three possible forms of $c_i(A_{ab}, x)$ on an arc, A_{ab}. In figure 3–3(a) the least cost route from input source node, N_i, to any point on A_{ab} is via node N_a; in figure 3–3(b) the least cost route from N_i to any point on A_{ab} is via node N_b; in figure 3–3(c) the least cost route changes along A_{ab}; for those points on A_{ab} between 0 and $\bar{x}$ the least cost route from N_i is via N_a and for those points between $\bar{x}$ and δ_{ab} it is via N_b. (Note that $c_i(A_{ab}, \bar{x}) = c_i(A_{ba}, \delta_{ab} - \bar{x})$.) Under the preceding assumptions the transport cost for an input or the output as a function of distance shipped *can* have convex regions, but on each arc the functions are concave; this is illustrated in figure 3–4.

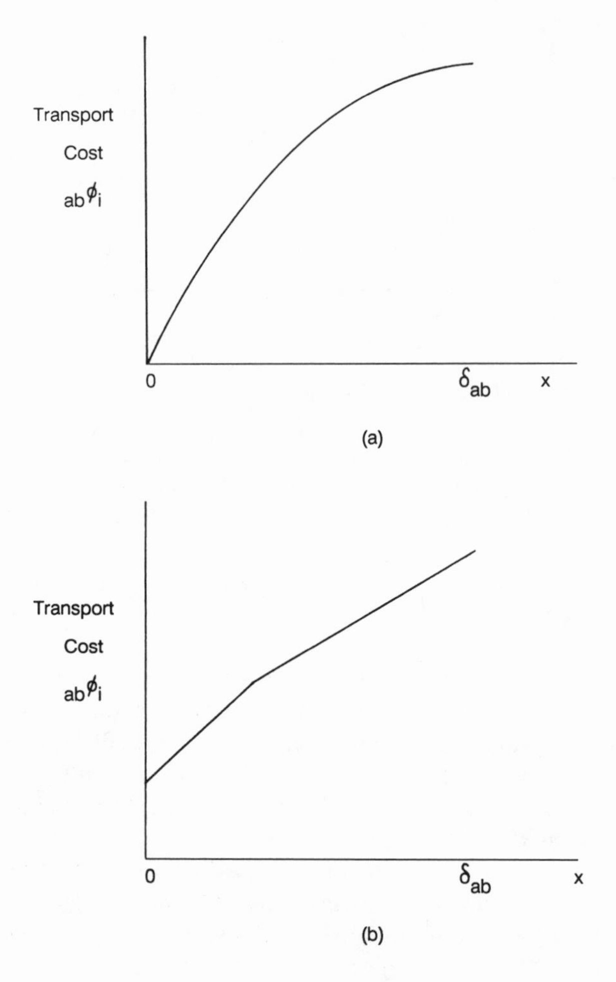

Figure 3–2. Typical Arc Transport Cost Functions, ${}_{ab}\phi_i$, on an Arc A_{ab}

Figure 3–3. Typical Transport Cost Functions, $c_i(A_{ab}, x)$, on an Arc A_{ab}

In (a) the least cost route from source i to every point on A_{ab} is through N_a; in (b) the least cost route to every point on A_{ab} is through N_b; in (c) the least cost route to all points on A_{ab} between 0 and $\bar{x}$ is through N_a, and to all points between $\bar{x}$ and δ_{ab} is through N_b.

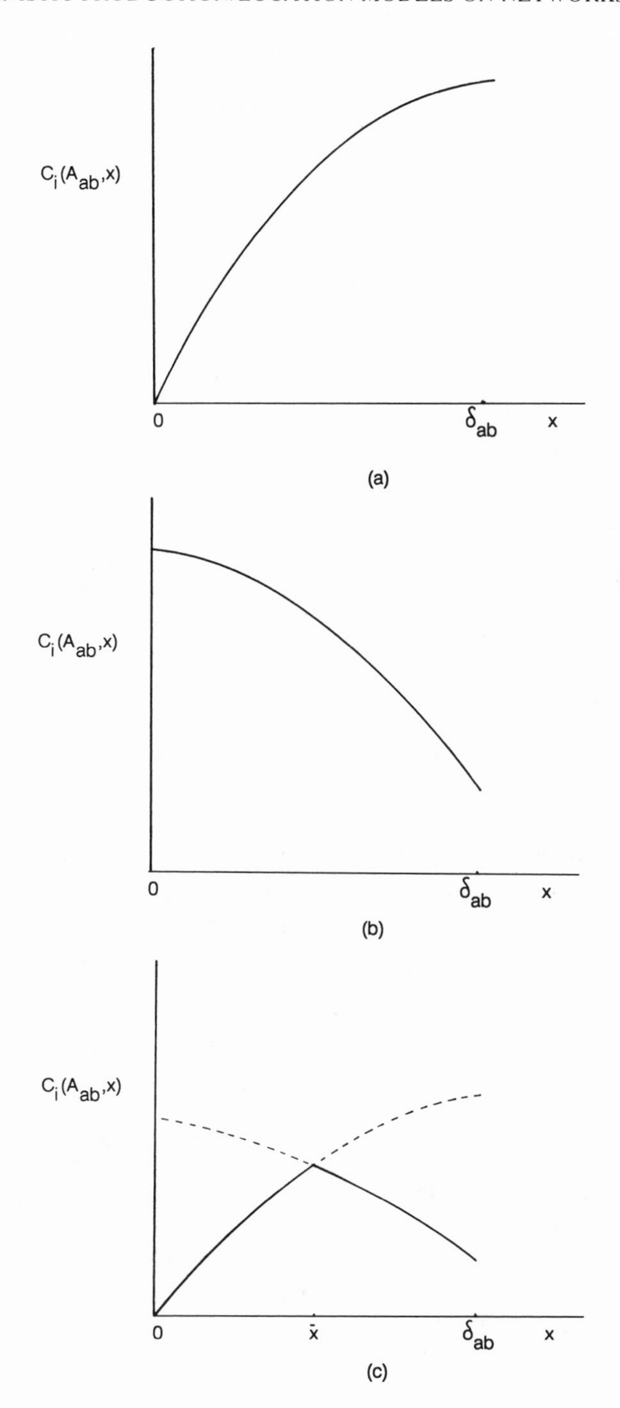
$C_i(A_{ab},x)$
0
δ_{ab}
x
(a)
$C_i(A_{ab},x)$
0
δ_{ab}
x
(b)
$C_i(A_{ab},x)$
0
$\bar{x}$
δ_{ab}
x
(c)

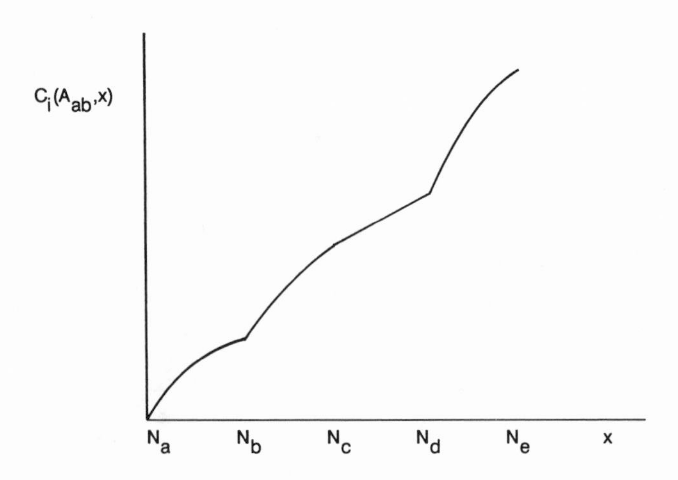

Figure 3–4. Transport Cost as a Function of Total Distance Across Several Arcs

For the transport cost conditions specified above, Wendell and Hurter [1973b] have developed node optimality results that are applicable to a wide variety of network location problems. They have shown that for any specified shipment pattern *if the transport cost functions are concave in distance on every arc*, then the following are true for the transport cost minimizing location.[3]

WH1: A node of η must be optimal.

WH2: If the transport cost function for any commodity shipped in positive amounts (input or output) is not linear in distance over *all* of A_{ab}, then the optimal location cannot be in the interior of A_{ab}.

WH3: A solution $(\mathbf{z}^*, x^*)$, where x^* is an interior location of A_{ab} (and $\mathbf{z}^*$ specifies the commodities shipped), can be optimal *only if* $(\mathbf{z}^*, x)$ is also optimal for *all* $x \in A_{ab}$.

In most pure network location problems the "shipping pattern," as defined by $\mathbf{z}^*$, is given a priori (e.g., the "median" problems considered by Hakimi [1964], Levy [1967], Minieka [1977] and many others). Wendell and Hurter's results apply as well to production/location and design/location problems where the optimal shipping pattern is determined in conjunction with choosing the optimal location. This is true because, as stated in chapter 1, for the optimal input and output levels chosen, the optimal plant location minimizes the total transport costs. Therefore the following result is immediate from WH1.

Theorem 3.1

In problems (PL3.1), (DL3.1), (PL3.2), and (DL3.2), if all transport cost functions ${}_{ab}\phi_i(x)$ and ${}_{ab}\phi_0(x)$ are concave in x on all arcs then there is a node of η which is an optimal location.

The main consequence of theorem 3.1 is that in searching for an optimum location, we need to consider only nodal locations; at each node we then solve a production subproblem. Further, from WH3 the existence of an interior optimum location is signaled by two adjacent nodes having the same optimum production solutions that result in the same minimum total cost. Although from a computational viewpoint theorem 3.1 is the crucial result, under certain conditions this result can be strengthened so as to exclude the possibility of interior (nonnodal) optimal locations, similar to theorems 2.1–2.5. The remaining theorems in this subsection identify the more important cases.

We first consider those models in which output transport costs are not included. To simplify the notation and analysis, and without loss of generality for this section, we will assume that the output price is fixed at p_0 rather than being a function of z_0. Define $(z_1^*, \ldots, z_n^*, A_{ab}^*, x^*)$ to be an optimal solution to (PL3.1) where $x^* \in A_{ab}^*$. If for any j, with $z_j^* > 0$, the transport cost function $c_j(A_{ab}^*, x)$ has the form in figure 3–3(c) (the shortest route from N_j to all $x \in A_{ab}$ is not via the same node), then $c_j(A_{ab}, x)$ is concave but not linear over all of A_{ab}. From WH2 it follows that x^* cannot be an interior point. (A direct argument is that if $c_j(A_{ab}, x)$ is concave but not linear, then the total transport cost function $TC(x) = \Sigma_{i=1}^n c_i(A_{ab}, x) z_i^*$ is concave and not linear over all of A_{ab} [Wendell and Hurter 1973b, lemma 2]. But the minimum of a nonlinear, concave function on A_{ab} can *only* occur at an endpoint of A_{ab} (i.e., a node). Therefore, we can restrict our attention to the case where the following property holds on the optimal arc, A_{ab}^*:

Property A. For each input i the least cost route from source node, N_i, to x is via N_a for *all* $x \in A_{ab}^*$ or via N_b for *all* $x \in A_{ab}^*$.

This property means that all the transport cost functions, $c_i(A_{ab}^*, x)$, look like those in figures 3–3(a) and 3–3(b), and not like the one in figure 3–3(c).

Now define M^a as the set of inputs with their least cost route to $x \in A_{ab}$ via N_a, and M^b as the set of inputs with their least cost route to $x \in A_{ab}$ via N_b. In this case, restricting attention to the arc A^*_{ab}, the Lagrangian for problem (PL3.1) is

$$L = p_0 F(z_1, \ldots, z_n) - \sum_{i=1}^{n} [p_i + c_i(A^*_{ab}, x)]z_i + \mu(\delta_{ab} - x)$$

and the first-order necessary conditions for an optimum are

$$\left.\begin{aligned} \partial L/\partial z_i &= p_0 F_i - p_i - c_i(A^*_{ab}, x^*) \leqslant 0 \quad &(3.3a)\\ z_i^*(\partial L/\partial z_i) &= 0 \quad &(3.3b)\end{aligned}\right\} \quad i = 1, \ldots, n$$

$$\partial L/\partial x = -\sum_{i \in M^a} {}_{ab}\phi_i'(x^*)z_i^* - \sum_{i \in M^b} {}_{ba}\phi_i'(\delta_{ab} - x^*)z_i^* - \mu^* \leqslant 0 \quad (3.4a)$$

$$x^*(\partial L/\partial x) = 0 \quad (3.4b)$$

$$\partial L/\partial \mu = \delta_{ab} - x^* \geqslant 0 \quad (3.5a)$$

$$\mu^*(\partial L/\partial \mu) = 0 \quad (3.5b)$$

$$\mu^* \geqslant 0 \quad (3.6)$$

where * denotes the optimal value of the variable. Condition (3.4a) identifies three cases for an optimal location on A_{ab}:

$$\text{If} \quad \begin{aligned} &\sum_{i \in M^a} {}_{ab}\phi_i'(x^*)z_i^* \\ &+ \sum_{i \in M^b} {}_{ba}\phi_i'(\delta_{ab} - x^*)z_i^* \end{aligned} \begin{cases} > 0 & \text{then } x^* = 0 = N_a \quad (3.7)\\ = 0 & \text{then } x^* = [0, \delta_{ab}] \quad (3.8)\\ < 0 & \text{then } x^* = \delta_{ab} = N_b \quad (3.9)\end{cases}$$

Conditions (3.7)–(3.9) are equivalent to (2.16)–(2.18) for the problem on a line, and can be developed similarly. The l.h.s. of (3.7)–(3.9) is the marginal transport cost of increasing x^* (moving the plant further from N_a, toward N_b); i.e., it is the net locational pull of the transport costs. Notice that condition (3.8) is equivalent (and can be used to prove) WH3. We can now prove the following.

Theorem 3.2

In (PL3.1) (and (DL3.1)) if all ${}_{ab}\phi_i$ are concave, then the optimal plant location can be *only* at a node of η.

Proof: We will prove the result by contradiction. Let $(z_1^*, \ldots, z_n^*, x^*)$ be an optimal solution to (PL3.1) where x^* is an interior point of some arc

A_{ab}. From WH2 and the previous discussion, it is sufficient to consider the case where all transport costs are linear on A_{ab} and property A holds; that is, ${}_{ab}\phi_i(x) = r_i x$ for every $i \in M^a$ and ${}_{ba}\phi_j(\delta_{ab} - x) = r_j(\delta_{ab} - x)$ for every $j \in M^b$, where $r_i, r_j > 0$ are constants. It is now sufficient to show that condition (3.8) cannot occur.

If (3.8) holds, then

$$\sum_{i \in M^a} {}_{ab}\phi_i'(x^*)z_i^* - \sum_{i \in M^b} {}_{ba}\phi_i'(\delta_{ab} - x^*)z_i^* = \sum_{i \in M^a} r_i z_i^* - \sum_{i \in M^b} r_i z_i^* = 0$$

So the optimal profit is

$$p_0 F(z_1^*, \ldots, z_n^*) - \sum_{i \in M^a} (p_i + r_i x^*)z_i^* - \sum_{i \in M^b} (p_i + (\delta_{ab} - x^*))z_i^*$$

$$= p_0 F(z_1^*, \ldots, z_n^*) - \sum_{i \in M^a} p_i z_i^* - \sum_{i \in M^b} (p_i + \delta_{ab})z_i^*$$

which is independent of the location variable, x. Thus, every location on the arc A_{ab} must be optimal with input vector $\mathbf{z}^* = (z_1^*, \ldots, z_n^*)$. But for every fixed location $\bar{x} \in [0, \delta_{ab}]$ the optimal input mix must satisfy the necessary condition

$$\frac{p_i + c_i(A_{ab}, 0) + r_i\bar{x}}{p_j + c_j(A_{ab}, 0) + r_j\delta_{ab} - r_j\bar{x}} = \frac{F_i(\mathbf{z}^*)}{F_j(\mathbf{z}^*)} \tag{3.10}$$

where $i \in M^a, j \in M^b, z_i^* > 0, z_j^* > 0$. (Notice that there must be at least one i and one j which satisfy these conditions, otherwise (3.8) could not hold and the optimal location would have to be an endpoint.) The r.h.s. of (3.10) is constant in $\bar{x}$, while the l.h.s. of (3.10) is strictly increasing in x. Thus, the equality can hold at most at one point on A_{ab}, not for all $\bar{x} \in [0, \delta_{ab}]$. But this contradicts WH3 and so (3.8) cannot hold. (The proof for (DL3.1) is identical.) □

This theorem demonstrates why the optimal location in example 3.1 *must* be at one of the nodes of the network. The only real difference between the network case and the models in chapter 2, is the availability of more inputs. In proving theorem 3.2 (and all node optimality theorems) we essentially prove endpoint optimality on a line segment, which happens to be the optimal arc. Consequently the proof of theorem 3.2 is similar to that of theorem 2.2. (Furthermore, the minimum cost function $C(x, z_0)$ is concave in x over A_{ab}, just as for location on a line segment.)

Even when output transport costs are included in the problem, as in (PL3.2), an interior location can be optimal only under very unusual conditions.

Theorem 3.3

In problems (PL3.2) and (DL3.2), if all ${}_{ab}\phi_i$ are concave, an interior location on an arc, A_{ab}, can be optimal only if (a) ${}_{ab}\phi_0$ and all ${}_{ab}\phi_i$, with $z_i^* > 0$, are linear over A_{ab}; (b) property A holds on A_{ab} for all input and output variables that are positive in value; (c) for *all* $z_i^* > 0$, the least cost route from N_i to every x on A_{ab} is via the same node (i.e., either M^a or M^b is empty); and (d) for some $z_{0m}^* > 0$ the least cost route from every x on A_{ab} to N_{n+m} is via N_b if M^b is empty and via N_a if M^a is empty.

Proof: (a) and (b) follow from WH2 and the argument preceding theorem 3.2.

(c) A necessary condition for an interior optimum (comparable to (3.8)) is

$$r_0\left(\sum_{m\in L^a} z_{0m}^* - \sum_{m\in L^b} z_{0m}^*\right) + \left(\sum_{i\in M^a} r_i z_i^* - \sum_{i\in M^b} r_i z_i^*\right) = 0 \qquad (3.11)$$

where L^a and L^b are the sets of output markets such that $z_{0m}^* > 0$ and their least cost routes from all $x \in A_{ab}$ to N_{n+m} are via N_a and N_b, respectively. Then the optimal profit can be shown to be independent of x, and if neither M^a nor M^b are empty, then an argument identical to that used in the proof of theorem 3.2 can be used to show the impossibility of (3.11) holding.

(d) If both M^a and L^a or M^b and L^b are empty, then either N_b or N_a must be optimal, respectively. So for an interior optimum to occur, some output must be sent to a market through the node other than that through which all the inputs are being received. □

Even if the conditions in theorem 3.3 are satisfied, for an interior optimum location to occur, the input and output transport costs must balance in a very special fashion. When (3.11) holds, the profit function is independent of x and if there is an interior optimum location, all points on A_{ab}^* are also optima. If, for example, M^a and L^b are empty as required by conditions (c) and (d), then

$$\sum_{m\in L^a} r_0 x z_{0m}^* = \sum_{i\in M^b} r_i(\delta_{ab} - x) z_i^*$$

must hold for all x. The delivered prices of the inputs, and thus the sum of the input transport costs, will change with x. For an interior optimum location to occur, the output transport costs must vary in *exactly* the same magnitude (but opposite direction). When all transport cost functions are linear on each arc, the chances of this occurring in any but the simplest cases, such as example 2.3, seem extremely small.

If any of the transport cost functions are not concave in distance shipped on the optimal arc, then an interior optimal location is possible (as illustrated by example 2.2). Even in such a case it is not uncommon for the optimum to be at a node.

Multifacility Models

Once we generalize the location space to be a network and allow n inputs and M output markets, the problem of concurrently locating more than one production facility becomes more relevant and worthy of study. Multifacility models are only meaningful if there are positive transport costs for the outputs. (If output transport costs are nill, it is usually optimal to locate all the plants at the same location unless there are limits on the availability of some inputs.) Therefore, we will formulate multifacility generalizations of (PL3.2) and (DL3.2) only. (Furthermore, we will *not* consider shipments between the plants as part of the firm's strategy.)

Suppose that the firm wishes to locate K plants on the network to supply the M customer markets. Using the previous notation and that used in chapter 1 for formulating (PL), let (A^k, x^k) designate the location of plant k on some arc A^k, z_{0m}^k be the amount of output produced at plant k and sent to market m, z_i^k be the amount of input i used at plant k—so $\Sigma_{m=1}^M z_{0m}^k = F(z_1^k, \ldots, z_n^k)$—and without loss of generality, let $p_{0m}(\Sigma_{k=1}^K z_{0m}^k) = p_{0m}$. The firm's problem is then to find the optimal locations, input mixes, and output rates for each plant and to allocate customers to plants:

$$\text{(PL3.3)} \quad \begin{aligned} &\underset{z_i^k, z_{0m}^k, x^k}{\text{maximize}} && \pi = \sum_{k=1}^{K} \left[\sum_{m=1}^{M} (p_{0m} - c_{0m}(A^k, x^k)) z_{0m}^k - \sum_{i=1}^{n} (p_i + c_i(A^k, x^k)) z_i^k \right] \\ &\text{subject to} && \sum_{m=1}^{M} z_{0m}^k = F(z_1^k, \ldots, z_n^k) \qquad \text{for } k = 1, \ldots, K \\ &&& z_{0m}^k \geqslant 0, \ z_i^k \geqslant 0 \qquad \text{for all } m, i, k \\ &&& x^k \in \eta \qquad \text{for all } k \end{aligned}$$

where the selection of x^k implies selection of an arc, A^k. If the plant output levels, $z_0^k = \Sigma_{m=1}^M z_{0m}^k$, are set in advance and customer demand levels, D_m, must be satisfied, then this reduces to the multifacility, design/location problem

$$\begin{aligned}
&\underset{z_i^k, z_{0m}^k, x^k}{\text{minimize}} && C = \sum_{k=1}^{K}\left[\sum_{m=1}^{M} c_{0m}(A^k, x^k) z_{0m}^k + \sum_{i=1}^{n} (p_i + c_i(A^k, x^k)) z_i^k\right] \\
&\text{subject to} && \bar{z}_0^k = \sum_{m=1}^{M} z_{0m}^k \\
&\text{(DL3.3)} && \quad = F(z_1^k, \ldots, z_m^k) \qquad \text{for } k = 1, \ldots, K \\
& && \sum_{k=1}^{K} z_{0m}^k = D_m \qquad \text{for all } m \\
& && z_i^k \geqslant 0 \qquad \text{for all } m, i, k \\
& && x^k \in \eta \qquad \text{for all } k
\end{aligned}$$

(Commonly only the total amount of output being sent to each market, D_m, is fixed, in which case the values z_0^k are variables, but $D_m = z_{0m} = \sum_{k=1}^{K} z_{0m}^k$. The subsequent results apply to this case as well; this modified form of (DL3.3) will be considered further in a later section.)

Wendell and Hurter's results apply to multifacility shipping patterns, as well as to single-facility patterns, so theorem 3.1 generalizes immediately to (PL3.3) and (DL3.3); that is, if all transport costs are concave on each arc, then (if an optimal solution exists) there is an optimal solution with all plant locations being nodes of the network. In fact, using the following inductive argument, it can be shown that all the previous node optimality results can be extended to comparable multifacility problems.

To see this consider an optimal solution to problem (PL3.3), say $(z_1^{1*}, \ldots, z_n^{K*}, z_{01}^{1*}, \ldots, z_{0M}^{K*}, x^{1*}, \ldots, x^{K*})$. Fixing the values of all the production and location variables for plants $2, \ldots, n$, the optimal location of plant 1 and the optimal values of the production variables for plant 1 must be optimal for the one-facility problem (PL3.2), otherwise a better solution to (PL3.3) could be found. Therefore, any node optimality properties that apply to (PL3.2) will apply to plant 1. The same reasoning can be used for plants $2, \ldots, n$, so that the node optimality properties of (PL3.2) apply to all K plants in (PL3.3). Therefore, we can conclude the following:

Theorem 3.4

Assuming that all transport costs are concave on every arc, an interior point of an arc, A_{ab}, can be an optimal plant location for problems (PL3.3) or (DL3.3) *only if* the conditions in theorem 3.3 hold for that plant and arc.

The preceding theorem justifies the attention given to discrete-set location problems, such as the fixed-charge problem presented in chapter 1 and other versions of location-allocation problems. Although an infinite number of plant locations are theoretically feasible for network location problems, this set usually can be reduced to a relatively small number of locations, which facilitates solution of the problem. Indeed, without node optimality properties, multifacility, network, production/location problems can be computationally intractable. This issue is explored in a later section.

Parametric Analysis

The network structure of the problems in this chapter makes it essentially impossible to characterize unambiguously how the optimal plant location(s) will change when problem parameters—such as output level, prices, and transport rates—change. For example, a change in input prices will often change the optimal plant location, but the change may not be to an adjacent node. The only apparent exceptions to this that have been developed formally are invariance results similar to those given by theorems 2.6 and 2.7. Thus, other than formally stating and briefly discussing these invariance results, in this section we will only be able to illustrate through a numerical example how parametric changes can alter a firm's optimal production and location solution.

For one-plant network problems, theorems 2.6 and 2.7 extend easily.

Theorem 3.5

In (DL3.1) let $(\mathbf{z}^1, x^1)$ be optimal for output level z_0^1. If F is homothetic, then for any output level z_0^2 there exists some scalar $\alpha > 0$ such that $(\alpha\mathbf{z}^1, x^1)$ is optimal. That is, the optimal input ratios and plant location are invariant to changes in the output level.

Theorem 3.6

In (DL3.2) let $(\mathbf{z}^1, x^1)$ be optimal for output level z_0^1. If F is homogeneous of degree one, then $((z_0^2/z_0^1)\mathbf{z}^1, x^1)$ is optimal for any output level z_0^2. That is, the optimal input levels change proportionately to the change in output and the optimal input ratios and plant location are invariant to changes in output level.

The proofs of these two theorems are essentially identical to those for theorems 2.6 and 2.7, respectively, and so they are omitted.

For the multiplant case, (DL3.3), the following holds.

Theorem 3.7

In (DL3.3) let $(Z^1, X^1) = (z_1^{11}, \ldots, z_n^{K1}, z_{01}^{11}, \ldots, z_{0M}^{K1}, x^{11}, \ldots, x^{K1})$ be optimal for plant output levels $Z_0^1 = z_0^{11}, \ldots, z_0^{K1}$ with a total cost of C^1. If F is homogeneous of degree one, then $(\alpha Z^1, X^1)$ is optimal for plant output levels αZ_0^1 with a total cost of αC^1, where $\alpha > 0$ is a scalar.

Proof: The proof is by contradiction. Suppose $(\alpha Z^1, X^1)$ is not optimal for plant output levels αZ_0^1. Instead let (Z, X) be optimal with a total cost of $C < \alpha C^1$. By the degree one homogeneity of F, this implies that $(Z/\alpha, X)$ is a feasible solution for (DL3.3) with output levels Z_0^1 and has a total cost of $C/\alpha < \alpha C^1/\alpha = C^1$. But this implies (Z^1, X^1) is not optimal for output levels (Z_0^1), a contradiction. Thus, $(\alpha Z^1, X^1)$ must be optimal for Z_0^1. □

Theorem 3.7 holds without change even if the total amount shipped to each market, z_{0m}, is fixed $(= D_m)$ rather than the amount produced at each plant, or if production functions differ between plants, as long as they are all degree-one homogeneous. If degree-one homogeneity does not hold, then the optimal solution (locations) can change as the output level changes. We demonstrate this latter fact and illustrate how sensitive network models are to parametric changes in the following example.

Example 3.2

Consider the network in figure 3–5. Suppose a firm produces one output from two physically distinct inputs, A and B, according to the production

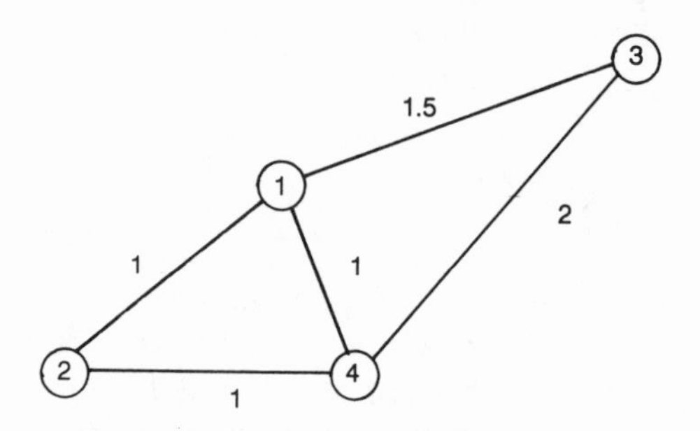

Figure 3–5. Network Location Space for Example 3.2

function: $F(z_A, z_B) = z_A^{0.6} z_B^{0.2}$. Input A is available at node 2 at a price of 5, and it is available at node 3 at a price of 4. Input B is available only from node 1 at a price of 3. (So we will refer to input B as input 1, input A from node 2 as input 2, and input A from node 3 as input 3.) The only market for output is at node 4. Suppose the cost per unit per unit distance to ship each of the inputs is 1, and the cost per unit per unit distance to ship the output is 2 (arc lengths are given in figure 3–5). The optimal solutions for three levels of output are given below for problem (DL3.2).

z_0	z_1^*	z_2^*	z_3^*	x^*	C^*
4	3.363	6.726	0	node 4	53.81
8	6.979	16.755	0	node 2	111.69
16	13.418	0	43.378	node 3	295.37

Notice how the optimal location for small levels of output is at the output market. Because of the decreasing economies of scale, as output increases the relative use of inputs to output increases so that the pull of the input markets gets stronger relative to that of the output market. At an output level of 8 the optimal location changes to node 2. As output increases further, to 16, the optimal location jumps again, to node 3. Observe that the optimal location jumps "over" the output market. Thus, locational (and production) changes to parametric perturbations are often not monotonic or smooth, in even a qualitative sense, on networks. This illustrates the difficulty with obtaining general statements regarding the sensitivity of the optimal solution. For network problems, frequently the only way to determine the effects of parametric changes is to resolve the problem for each set of parameter values. The issue of solution methods is treated next.

Solution Methods and Their Relationship to Other Location Models

Single-Plant Problems

In practice, transportation cost rates are usually concave in distance shipped, so the node optimality theorems given above apply. As a result any solution method can restrict its locational search to the nodes of the network. Ideally we would like to find an algorithm that starts at some node and progresses to other nodes producing monotonically improving objective function values, and that eventually converges to the optimum without having to evaluate every node explicitly (i.e., solve a production subproblem for every node). Unfortunately, no such algorithm, which will find the optimum in general, has been found. The only procedure that can guarantee convergence to the optimum for general problems is a total nodal enumeration. This is a computationally burdensome approach for networks of large size because the production subproblem at every node must be solved.

For many problems, however, this enumerative method can be simplified and improved without jeopardizing convergence to the optimum. The simplifications include (a) the elimination of some nodes from consideration without having to solve a production subproblem for the node, and (b) making the solution of the individual production subproblems more efficient. These simplifications are based upon the following observations.

1. If there is no limitation on the amount of input that can be supplied from any source location, then it would never be advantageous to buy the *same physical input* from more than one source location; specifically, only the lowest delivered cost supplier of the input needs to be used. Thus, if there are five physically different inputs, but they are available from twenty different source locations, then only five production variables (rather than twenty) need to be included in the production subproblem at any node. In example 3.1, at nodes 1 and 2 only the variables z_1 and z_2 are included in the production subproblem; at node 3 only the variables z_2 and z_3 are included, and at node 4 only the variables z_1 and z_4 are included.

2. If an iterative numerical algorithm is used to determine the production optimum at a node (rather than closed-form solution of first-order conditions), then the production solution at one node can be used as a starting point for finding the production solution at an adjacent node (where the values are carried over the same *physical* inputs).

3. If the lowest delivered price for each physically distinct input at node N_a is at least as large as at node N_b (and the selling price of output minus transport costs is no larger), then the best solution at N_b must be at least as good as the best solution at N_a, and so node N_a can be eliminated from consideration without solving the production subproblem at N_a. In example 3.1 node 4 can be eliminated by this property because the lowest delivered prices for inputs A and B are 3.9 and 4.5, respectively, while at node 2 they are 3.8 and 4.4, respectively.

Greater efficiencies in the algorithm do not seem possible without jeopardizing optimality; however, it is possible to construct heuristics that will find good (though not always optimal) solutions, with limited node search. For brevity, discussion of such heuristics is postponed until the next subsection.

Transport costs are usually concave in distance and in quantity shipped. However, in the presence of congestion or other aberations transport costs may not be concave, so that an interior point of an arc can be an optimal location. Without nodal optimality, the solution process becomes substantially more difficult. Under these conditions the optimal location on each arc first has to be determined either explicitly (using some numerical search, such as a grid-search method similar to that proposed by Nijkamp and Paelinck [1973] for planar problems (see chapter 4) or a standard nonlinear programming algorithm), or implicitly by eliminating all interior locations using some theorem similar to theorems 3.1–3.4. For general transport cost functions a priori elimination of interior points on the arc is unlikely unless very special conditions exist. Consequently, if node optimality cannot be proved, heuristics often provide the best practical approach; these are discussed below.

Multiplant Problems

Multifacility location models have received considerable attention from operations researchers, management scientists, and transportation scientists. (See Handler and Mirchandani [1979] for a good discussion of network location models and their solution methods.) The models considered by those researchers are special cases of the production/location models presented in this book. Those models focus primarily on the location variables and either ignore the production variables completely or assume that a separate set of production optimizations has been performed as a preparatory step. Although the enhanced ability to solve models using standard techniques does not necessarily justify the modeling errors that result from discarding the production/location approach, in practice, these approaches are often necessary in order to solve multifacility problems. In spite of their possible suboptimization for the underlying production/location problem, studying such approaches is instructive and their use is often appropriate.

The types of models most relevant for out purposes here are those that have been referred to in the literature as "warehouse location" or "fixed-charge location" models. We begin by assuming that the transport cost structure (or some other aspect of the problem) allows us to reduce the set of feasible plant locations to a finite set, say the nodes of the network. At each node a production (or design) "preanalysis" must then be performed. Specifically, at each feasible site (node), j, the firm's cost (or profit) function, say $C_j(z_0^j)$, must be derived. The firm then must choose the set of locations (and the corresponding production solution) that minimizes total cost (or maximizes profit, depending on the form of the problem).

If the firm's production function is of a well-behaved form (e.g., a Cobb-Douglas production function), and the number of physically distinct inputs is small, derivation of the plant's cost or profit function at each node is reasonably manageable. The assumptions made in the previous chapters have implied that the resulting cost functions should be continuous and differentiable. We do not require these properties here, and, following the tradition of the fixed-charge problem literature, we now assume that the cost functions take the form:

$$C_j(z_0^j) = \begin{cases} 0 & \text{if } z_0^j = 0 \\ f_j + h_j(z_0^j) & \text{if } z_0^j > 0 \end{cases}$$

where f_j is a fixed cost of opening the plant at site j and h_j is a variable cost function which depends on output level. Included in the variable cost function, h_j, are the costs of transporting inputs to site j. Thus, the only transport cost explicitly included in these models is the cost of transporting output to markets. If the firm's production function is homogeneous of degree one (after the fixed cost of opening the plant), then the variable cost function reduces to a linear form: $h_j(z_0^j) = g_j z_0^j$, where $g_j > 0$ is a constant (this assumes all input prices and transport rates are constant). For now we will assume this form of cost function.

To make the model precise, suppose there are M spatially distinct customers (located at nodes of the network), with customer m demanding D_m units of output no matter what its cost. Let there be J nodes that are possible plant sites, and let S_j be the maximum output level possible at site j (this can be infinite). For each possible plant site a cost function $C_j(z_0^j)$ as described above is assumed to have been derived. Finally, r_{jm} is the (constant) per unit cost of transporting output from plant site j to customer m. The firm's problem can now be formulated as the mixed-integer program, (FC), given in chapter 1.

$$\text{Let} \quad y_j = \begin{cases} 0 & \text{if a plant is not opened at site } j \\ 1 & \text{if a plant is opened at site } j \end{cases}$$

for $j = 1, \ldots, J$, and let z_{0m}^j be the number of units of output produced at site j and shipped to customer m. Assuming customer demands must be satisfied, the firm's problem is then to

$$\text{minimize} \qquad C_{\text{FC}} = \sum_{j=1}^{J} \left[f_j y_j + \sum_{m=1}^{M} (g_j + r_{jm}) z_{0m}^j \right] \qquad (3.12)$$

(FC)

$$\text{subject to} \quad \sum_{j=0}^{J} z_{0m}^{j} = D_m \qquad m = 1, \ldots, M \tag{3.13}$$

$$\sum_{m=1}^{M} z_{0m}^{j} \leq S_j y_j \qquad j = 1, \ldots, J \tag{3.14}$$

$$z_{0m}^{j} \geq 0 \tag{3.15}$$

$$y_j \in \{0, 1\} \tag{3.16}$$

Notice that constraints (3.14) ensure that z_{0m}^{j} will equal zero for all m if y_j equals zero (i.e., no plant is opened at site j). Numerous variations and modifications of (FC) have been presented in the literature. Before discussing some of them and their advantages or disadvantages, we will illustrate the general approach used to solve these problems.

As mentioned in chapter 1, (FC) is a 0–1 mixed-integer program. Almost all of the methods proposed for its solution are based upon a branch-and-bound approach. The branch-and-bound strategy is based upon the following ideas:

1. Suppose (Y^a, Z_0^a) is a feasible solution for (FC) and it has a total cost of $C_{FC}(Y^a, Z_0^a)$; then $C_{FC}(Y^a, Z_0^a)$ is an upper bound (UB) on the minimum possible total cost for (FC).
2. We can divide the set of possible solutions into subsets by fixing the value of one or more of the y_j's; e.g. if we fix $y_1 = 0$, then we are "defining" the subset of solutions that have no plant at site 1; if we set $y_1 = 1$ and $y_2 = 0$, then we are "defining" the set of solutions that have a plant at site 1 but not at site 2.
3. Suppose we have a relatively easy way to obtain a lower bound, $LB(S)$, on the total cost of the best feasible solution in some subset, S, of solutions; e.g., $LB(y_1 = 1)$ is a lower bound on the total cost of all solutions that have a plant opened at site 1.
4. If $UB \leq LB(y_1 = 1)$, then we can eliminate from consideration all solutions that have a plant opened at site 1 (the usual terminology is that we "fathom" the set defined by $y_1 = 1$). The effectiveness of this approach clearly depends greatly upon obtaining a good feasible solution quickly (i.e., with as small an upper bound as possible), and with devising a way of obtaining "tight" lower bounds on subsets (i.e., with values as close to the true minimum cost for solutions in that subset) with a minimum of computation. The various branch-and-bound solution methods that have been proposed differ mainly in how they obtain (and improve upon) good

feasible solutions, how they select subsets of solutions to evaluate, and especially how they obtain lower bounds for the best solution in the subsets.

We will illustrate this general approach using a simple branch-and-bound method; more efficient methods will be discussed later.

Example 3.3 (based upon an example by Khumawala [1972]

Suppose a firm faces a form of problem (DL3.3) on a network. The firm has eight spatially distinct customers, and it has decided in advance that it will (must) supply D_m units of output to each customer m. From the transport cost structure the firm has deduced that the optimal plant locations must be at nodes of the network. Furthermore, for legal, technological, and other reasons the firm has been able to eliminate all but five nodes as potential plant sites. At each node the firm then derives the cost of opening and operating a plant at that node. The cost and demand data are given in table 3–1. (A cost of 999 is imposed where transport of the output between site j and customer m is impossible.) To simplify the illustration, we assume that the capacity at each site is 50 (so there is essentially no limit on output from any one site, i.e., we have an "uncapacitated" fixed-charge problem). Thus, the original problem can now be put into the form of (FC).

We will now solve (FC) for the data in this example. Suppose we begin by choosing the following feasible solution: $y_4 = 1$, all other $y_j = 0$, and $z^4_{0m} = D_m$ for all m and $z^j_{0m} = 0$ for all $j = 1, 2, 3, 5$; $m = 1, \ldots, 8$. The total cost for this solution is 1595 (notice that 1595 is an upper bound—

Table 3–1. Cost and Demand Data for Example 3.3
costs $g_j + r_{jm}$

Plant Site j	*Customer m* 1	2	3	4	5	6	7	8	*Fixed Cost* f_j
1	12	90	50	999	60	999	18	999	200
2	21	999	75	48	55	14	11	33	200
3	18	95	55	39	50	999	999	39	200
4	21	95	75	36	65	8	16	24	400
5	17	75	55	30	70	13	20	999	300
Demand D_m	10	2	2	5	1	15	10	5	

UB—on the lowest total cost for the problem). We now divide the set of possible solutions into two sets: those that have $y_1 = 1$ (a plant open at site 1) and those that have $y_1 = 0$ (no plant at site 1). We can get a lower bound on the best solution to (FC) with $y_1 = 1$ by solving the linear programming "relaxation" of (FC) with $y_1 = 1$; that is, we set $y_1 = 1$ in the problem and replace constraints (3.16) with

$$0 \leqslant y_j \leqslant 1 \qquad j = 2, 3, 4, 5$$

The optimal solution to the LP relaxation is $y_1 = 1$, $y_2 = 0.20$, $y_3 = 0.02$, $y_4 = 0.40$, $y_5 = 0.14$, customers 1 and 3 are supplied by site 1, 7 is supplied by site 2, 5 by 3, 6 and 8 by 4, and 2 and 4 by 5. The resulting cost of 1366 is a lower bound on the solution with site 1 "open." Because $LB(y_1 = 1) < 1595$ we cannot delete the set of solutions with $y_1 = 1$ from consideration. If we repeat this for the set with $y_1 = 0$, we get $y_1 = 0$, $y_2 = 0.20$, $y_3 = 0.26$, $y_4 = 0.40$, $y_5 = 0.14$, with the same customer supply pattern as for $y_1 = 1$ except site 3 takes over the customers served by site 1, and $LB(y_1 = 0) = 1284$. Again, because $LB(y_1 = 0) < UB$, we cannot fathom this set of solutions.

Therefore, we branch again by dividing one of the previous subsets into smaller subsets. Suppose we select the set with $y_1 = 1$ (in practice, we would not choose this subset to branch on because its lower bound is larger than for the other available branch, but we are choosing this for illustrative purposes); we divide this into the sets: $(y_1 = 1, y_2 = 1)$ and $(y_1 = 1, y_2 = 0)$. If we solve the LP relaxation for the first case, we get the solution $y_1 = 1$, $y_2 = 1$, $y_3 = 0.02$, $y_4 = 0.10$, $y_5 = 0.14$, customers 1 and 3 are served by site 1, 6 and 7 by 2, 5 by 3, 8 by 4, and 2 and 4 by 5; and $LB(y_1 = 1, y_2 = 1) = 1496 < UB = 1595$ so this set cannot be fathomed. Repeating this for the other subset gives $LB(y_1 = 1, y_2 = 0) = 1396 < UB$, so no fathoming can occur.

We now select the subset $(y_1 = 1, y_2 = 1)$(again, in practice we would not choose this subset) and divide this into smaller subsets by fixing $y_3 = 1$ and $y_3 = 0$. Solving the LP relaxations for each gives: $LB(y_1 = 1, y_2 = 1, y_3 = 1) = 1692 > UB = 1595$, so we can eliminate all solutions with plants open at sites 1, 2, and 3; and $LB(y_1 = 1, y_2 = 1, y_3 = 0)1497 < UB$, so we cannot fathom this subset. If we take this latter subset and branch on it by fixing the value of y_4 and solving the corresponding LP relaxations we get: $LB(y_1 = 1, y_2 = 1, y_3 = 0, y_4 = 1) = 1767 > UB$, so we can fathom all solutions in this subset; and $LB(y_1 = 1, y_2 = 1, y_3 = 0, y_4 = 0) = 1502 < UB$, so this subset cannot be fathomed. Branching on this latter subset gives us the *feasible* solutions to (FC) of $(y_1 = 1, y_2 = 1, y_3 = 0, y_4 = 0, y_5 = 1)$ with $LB = 1745 > UB$ and $(y_1 = 1, y_2 = 1, y_3 = 0,$

$y_4 = 0, y_5 = 0$) with $LB = 1580$. Because this latter solution is feasible and better than our current UB solution, we update our UB to 1580 and keep this latter solution as our "incumbent" solution.

We can now backtrack to the subset of solutions ($y_1 = 1, y_2 = 0$), which has not been fathomed; and we branch on it by fixing y_3. Repeating this process generates a branch-and-bound "tree," the upper part of which is given in figure 3–6. The optimal solution is to open plants at sites 1 and 2 and supply customers 1 and 3 from site 1 and the other customers from plant site 2. The minimum total cost is 1580.

The branch-and-bound method will eventually converge to an optimum,

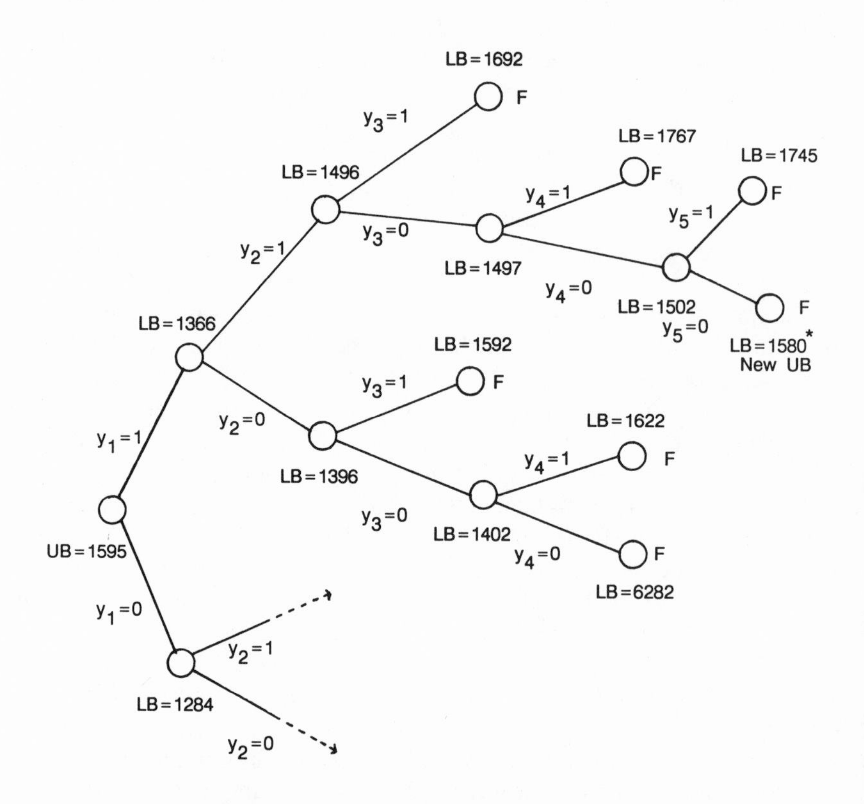

Figure 3–6. Upper Portion of the Branch-and-Bound Tree for Example 3.3

but the process is very time consuming if the bounding method is not good at producting tight bounds and the selection of branching variables is not done efficiently. In fact, many ways have been developed for improving the branching and bounding phases of the algorithm. We will now briefly review the most promising approaches; we encourage the reader to see Francis, McGinnis, and White [1983] or Krarup and Pruzan [1983] for more complete reviews of the relevant literature.

Because most of the algorithmic work developed has been for the "uncapacitated" fixed-charge, we will begin with that case. In addition, it is often convenient to define the z^j_{0m} as the fraction of customer m's demand supplied by the plant at site j. Because there is no limit on the amount of output that a plant can produce, constraints (3.14) no longer apply. However, it is necessary to have some way of ensuring that no output is produced at a site unless a plant is opened there, i.e., $z^j_{0m} \leq y_j$. The earliest algorithmic work on this problem—by Efroymson and Ray [1966] and Khumawala [1972]—used the formulation

(WUFC)

$$\text{minimize} \qquad C_{\text{FC}} = \sum_{j=1}^{J} \left[f_j y_j + \sum_{m=1}^{M} (g_j + r_{jm}) z^j_{0m} \right] \tag{3.17}$$

$$\text{subject to} \qquad \sum_{j=1}^{J} z^j_{0m} = 1, \qquad m = 1, \ldots, M \tag{3.18}$$

$$\sum_{m=1}^{M} z^j_{0m} \leq M y_j \qquad j = 1, \ldots, J \tag{3.19}$$

$$0 \leq z^j_{0m} \leq 1 \qquad \text{for all } j, m \tag{3.20}$$

$$y_j \in \{0, 1\} \qquad j = 1, \ldots, J \tag{3.21}$$

where the g_j and r_{jm} are rescaled accordingly. An alternative formulation is to replace (3.19) with the constraints

$$z^j_{0m} \leq y_j \qquad \text{for all } j, m \tag{3.22}$$

The formulation that includes (3.22) will be designated (SUFC). Although the two formulations are equivalent in terms of integer programming solutions, their linear programming relaxations are quite different. Specifically, feasible solutions for the relaxation of (SUFC) are always feasible for the relaxation of (WUFC), but the opposite is not true. For this reason (WUFC) is called the "weak" formulation of the uncapacitated fixed-charge problem while (SUFC) is called the "strong" version.

Formulation (WUFC) has far fewer constraints than (SUFC), and its LP relaxation is trivial to solve. However, the bounds obtained from this relaxation are usually not as good as for (SUFC), and rarely does the relaxation produce integer solutons, unlike (SUFC) which often produces integer solutions for the LP relaxation. Schrage [1975] first noticed the advantages of using (SUFC) rather than (WUFC), and since that time all major algorithmic work has been based upon that form. Erlenkotter [1978] proposed a branch-and-bound method that uses the dual of a relaxation of (SUFC) to compute lower bounds. Not only are the bounds relatively tight, but they can be obtained from feasible dual solutions that are simple to identify. (The superiority of Erlenkotter's method vis a vis the method used in example 3.3 is demonstrated by the fact that his approach requires explicit computation of lower bounds for only a couple of subsets whereas the method used above required over 20 bounds to be computed. For larger problems the relative performance of Erlenkotter's method is even better.)

If capacity restrictions do exist then (FC) can be formulated in either a weak form (WFC), which is (WUFC) with (3.19) replaced by

$$\sum_{m=1}^{M} D_m z_{0m}^{j} \leqslant S_j y_j \qquad j = 1, \ldots, j \tag{3.23}$$

or a strong form (SFC) which is (WFC) with (3.22) replacing (3.20).

The most successful solution methods for this problem, such as those of Geoffrion and McBride [1978], Nauss [1978], and van Roy [1986], use Lagrangean relaxation to obtain lower bounds. The general idea of these approaches is that for a relaxed LP some of the constraints are "dualized" and taken into the objective function along with Lagrangean multipliers. A feasible solution to this Lagrangean relaxation establishes a lower bound. The success of the method results from the fact that the Lagrangean problems are chosen so as to be easy to solve and produce very strong lower bounds.

In each of the preceding problems the maximum number of plants to be located is not fixed—the optimum number is implicitly determined within the solution. If an upper limit of, say K, is imposed on the number of facilities that can be opened, the constraint

$$\sum_{j=1}^{J} y_j \leqslant K \tag{3.24}$$

would have to be imposed. (Recall that the multifacility design/location problem, (DL3.3), was defined with the number of facilities to be located fixed at K.) The resulting problem is usually referred to as the "generalized

K-median problem"; we will designate it by (GKM). This problem is computationally more difficult to solve than the previous problems, and in some cases the methods suggested for the normal fixed-charge problems break down. A number of approaches have been proposed and tested, however. Probably the most extensive study on this problem is by Cornuejols, Fisher, and Nemhauser [1977]. Their work includes both heuristics and optimization methods.

All of the preceding models assume that the marginal costs are constant. When economies of scale exist in either production or transportation the problem becomes even more complex. The computational work on these problems is limited, but Efroymson and Ray [1966], Rech and Barton [1970], and Soland [1974] have suggested methods for these cases.

Although considerable progress has been made in the past decade on solution methods for location problems, it should be kept in mind that the effectiveness of many algorithms depends crucially on the problem structure and assumptions. The addition of what appear to be simple side constraints can make problems go from easily solvable to intractable. In those cases heuristics are often the only practical way of "solving" the problem. With this thought we return to a more direct consideration of production/location.

In most real situations network production/location problems can be reduced to discrete set problems (i.e., restricted to a subset of the nodes), and derivation of the cost (profit) functions at each relevant node is tractable. In those cases reduction of the production/location problem to some variation of the fixed-charge problems is appropriate, and it allows us to utilize efficient optimization algorithms. (Solution is especially easy in the special case when the marginal production costs are constant, such as in (3.12); but as mentioned above, solution methods have also been developed for the case where marginal production costs are decreasing.) In those cases when one or both of these conditions is violated, and especially if the underlying network is large, optimization methods are usually intractable. One heuristic for planar problems, proposed by Cooper [1964, 1967], can be applied to network problems as well. We will briefly present the idea underlying the method now and give more detail for planar problems where it has received more attention.

The underlying thesis of this book is that the production and location decisions are interrelated and cannot be separated without risking suboptimization unless special conditions hold. A reasonable heuristic, however, is to solve these two problems separately in an iterative fashion. For example, consider problem (DL3.3). We will use the notation (Z, X) to designate a feasible solution to (DL3.3). We begin by choosing a set of plant locations, say X^1 (this can be done arbitrarily or by using some information about the problem). With these plant locations fixed, we now solve (DL3.3) in terms of the production variables; that is, we find the optimal production plan for these plant locations, call it Z^1. Now we fix the production variables in (DL3.3) at values Z^1 and solve the remaining subproblem in terms of the location variables, obtaining a set of location X^2. We then fix the locations at X^2 and solve (DL3.3) for the production variables. We repeat this until the location and production solutions remain the same on successive iterations.

The advantage of this approach is that each subproblem, one in terms of only the production variables and the other in terms of only the location variables, is much easier to solve than the combined production/location problem. In addition, the algorithm has desirable monotonicity and convergence properties (e.g., the solutions on successive iterations cannot get worse; see Wendell and Hurter [1976] and Wendell and Rosenblum [1980] for a discussion of the properties of this type of algorithm). The main drawback is that it is not guaranteed to find a global optimum to the original problem. The effectiveness of this type of heuristic has not been investigated for network production/location problems and is a potential area for research.

Summary and Conclusions

The models in the chapter are more realistic and generally applicable than those presented in chapter 2 because of the generality of the network "location space." The cost of this additional applicability is significantly greater computational complexity and less informative parametric results. In spite of these limitations real production/location problems have been, and will continue to be, solved in practice. For the most part these have been solved as fixed-charge type location problems.

A primary goal of this chapter has been to clarify the relationship between general production/location formulations of the firm's problem and those formulations used in practice. The node optimality results in the first two sections of the chapter justify reduction of network problems to discrete set problems of the fixed-charge form. In addition, the homotheticity and homogeneity properties of the production function, which make optimal solutions invariant to proportional output (demand) perturbations, also make tractable derivation of a cost function for each node site. Thus, the reduction of general production/location problems to simpler location problems is often justified, and it allows us to take advantage of the significant algorithm developments that have occurred in the past two decades for solving these discrete set problems.

Notes

1. Different terminology for the same problem is employed in the operations research literature than is common in regional science. For example:

> The *minisum problem* is that of locating one facility so as to minimize the weighted sum of distance traveled. In Euclidean space this is the *Weber problem*.
>
> The *minimax problem* is that of locating one facility so as to minimize the maximum distance between the facility and any demand point.
>
> The *K-median problem* is that of locating *K* facilities so as to minimize the weighted sum of all distances traveled. It is the multifacility Weber problem.
>
> The *K*-center problem is that of locating *K* facilities so as to minimize the maximum distance from the closest facility to any demand point.

2. Least cost routes are assumed to be known; see Lawler [1976] for methods of finding least cost routes.

3. A specified shipment pattern means that fixed quantities of commodities are to be shipped between given points and a location to be chosen.

4 DETERMINISTIC PRODUCTION/LOCATION MODELS ON A PLANE

While firm-level applications often involve location on a network, models in which the firm can locate anywhere within a subset of a plane can be useful and important in several ways. Planar models may satisfactorily represent real situations when the underlying transport network is dense, or when transport costs are relatively proportional to Euclidean distances (e.g., when transport is performed by air or sea using company fleets). In addition, planar models permit more interesting, informative, and formal sensitivity analyses than do network models and are, accordingly, an important object of study.

Although optimality conditions can be identified for general n-input, planar models, most related research has dealt with simpler two- and three-input models because of their greater tractability. Consequently, we begin this chapter by formulating a two-input, triangular space model. We then generalize this model to the n-input case and include multifacility, planar production/location models, which although conceptually plausible, have received little research attention. Extensive sensitivity analyses are then performed for these planar models. This is followed by a discussion of possible solution methods that are illustrated with numerical examples.

A Triangular Two-input Model

Consider a firm that produces one output using two spatially distinct inputs. The input sources are located in a plane at points (a_1, b_1) and

(a_2, b_2); the output is shipped to a market located at point (a_0, b_0). Without loss of generality and following standard convention, we assume that the per unit cost of transporting input i from (a_i, b_i) to plant location (x, y), is $c_i(x, y) = r_i d_i(x, y)$ where $d_i = [(a_i - x)^2 + (b_i - y)^2]^{1/2}$ is the Euclidean distance between (x, y) and (a_i, b_i), and $r_i > 0$ is the (constant) unit transport cost per unit distance; that is, the per unit transport cost is proportional to Euclidean distance transported.[1] Let $c_0(x, y)$ be defined equivalently to represent the per unit cost of shipping output from the plant to (a_0, b_0). (See figure 4–1.) Letting all other notation be the same as that used in chapters 1–3, and for simplicity, letting $p_0(z_0) = p_0$, the firm's problem can be written as

$$\text{(PL4.1)} \quad \underset{z_i, x, y}{\text{maximize}} \quad \pi = [p_0 - r_0 d_0(x, y)]F(z_1, z_2) - \sum_{i=1}^{2} [p_i + r_i d_i(x, y)]z_i$$

$$\text{subject to} \quad z_i \geqslant 0 \qquad i = 1, 2 \tag{4.1}$$

$$(x, y) \in R^2 \tag{4.2}$$

When the output level is fixed at $\bar{z}_0$ a priori, (PL4.1) becomes the design/location problem

$$\text{(DL4.1)} \quad \underset{z_i, x, y}{\text{minimize}} \quad C = r_0 d_0(x, y)\bar{z}_0 + \sum_{i=1}^{2} [p_i + r_i d_i(x, y)]z_i$$

$$\text{subject to} \quad \bar{z}_0 = F(z_1, z_2)$$

$$z_i \geqslant 0 \qquad i = 1, 2 \tag{4.1}$$

$$(x, y) \in R^2 \tag{4.2}$$

Our investigation of these problems will be aided by exploring a numerical example by Nijkamp and Paelinck [1973].

Example 4.1

For problem (DL4.1), let $(a_0, b_0) = (0, 0)$, $(a_1, b_1) = (10, 0)$, and $(a_2, b_2) = (0, 10)$. The firm's production function (which is degree-one homogeneous) is given by $z_0 = z_1^{0.4} z_2^{0.6}$, and the output level is fixed at $\bar{z}_0 = 100$. Finally, let the transport cost rates be $r_0 = r_1 = r_2 = 1$, and let the prices of the output and inputs, at their markets, be $p_0 = 100$, $p_1 = 20$, $p_2 = 20$. The optimal solution for this problem is to locate at $(1.547, 3.352)$, $z_1^* =$

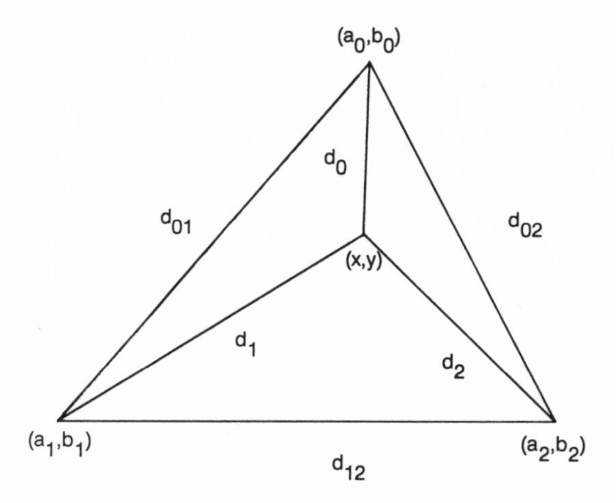

Figure 4–1. Triangular Location Space for Problems (PL4.1) and (DL4.1)

74.68, $z_2^* = 121.50$; the maximum profit is 4199.16; and the total cost is 5800.84.

It is straightforward to prove (see Wendell and Hurter [1973a]) that the optimal plant location for (PL4.1) and (DL4.1) must be in the convex hull of the set $\{(a_0, b_0), (a_1, b_1), (a_2, b_2)\}$; that is, in the triangle formed by the market and source locations.[2] Nijkamp and Paelinck (1973) present the results for over twenty numerical examples of (DL4.1). In all but two cases the optimal location is at a corner of the triangle (market or source). In the remaining two cases (including example 4.1 above) the location was in the interior of the triangle; in no case was the optimal location on an edge of the triangle other than at a corner. A reasonable question is whether or not these numerical results hold more generally or are simply a coincidence. This issue was addressed by Kusumoto [1984], who characterized when an interior or boundary location can or cannot be optimal.[3] Kusumoto, as well as others, used a geometric approach based upon a polar coordinate representation of the location space.

Consider problem (DL4.1), where the locations are expressed in the previously defined rectangular coordinate system. For simplicity, we will drop the argument (x, y) and let d_i be the Euclidean distance from the plant location to (a_i, b_i); d_{ij} will be the Euclidean distance from (a_i, b_i) to (a_j, b_j). The Lagrangian for this problem is then

$$L = r_0 d_0 \bar{z}_0 + \sum_{i=1}^{2} (p_i + r_i d_i) z_i + \lambda[\bar{z}_0 - F(z_1, z_2)]$$

and the first-order necessary conditions are

$$\left.\begin{aligned} \partial L/\partial z_i &= p_i + r_i d_i - \lambda F_i \geqslant 0 \\ (\partial L/\partial z_i) z_i^* &= 0 \end{aligned}\right\} \quad i = 1, 2 \qquad \begin{aligned}(4.3\text{a})\\(4.3\text{b})\end{aligned}$$

$$\partial L/\partial x = r_0 \bar{z}_0 d_0^x + \sum_{i=1}^{2} r_i z_i^* d_i^x = 0 \tag{4.4}$$

$$\partial L/\partial y = r_0 \bar{z}_0 d_0^y + \sum_{i=1}^{2} r_i z_i^* d_i^y = 0 \tag{4.5}$$

$$\partial L/\partial \lambda = \bar{z}_0 - F(z_1^*, z_2^*) = 0 \tag{4.6}$$

$$\lambda^* \geqslant 0, \ z_i^* \geqslant 0$$

where $d_i^x = (x - a_i)/d_i$ and $d_i^y = (y - b_i)/d_i$, $i = 0, 1, 2$, are the partial derivatives of d_i with respect to x and y, respectively.[4] (Notice that d_i^x and d_i^y are undefined at (a_i, b_i); this issue is discussed later.)

When (x, y) is given, (DL4.1) has an objective function that is linear in the input variables, z_i, and constrained by the input combinations that are feasible to produce $\bar{z}_0$ output. When the production function is quasiconcave, the isoquants are convex and the set of feasible input combinations for $\bar{z}_0$ is convex. On the other hand, for a given input vector, (DL4.1) has an objective function that is convex in x and y because of our choice of an L_p distance measure—the Euclidean. Thus, the imbedded production and location subproblems are both convex programs. However, when the z_i's and (x, y) are variables to be determined simultaneously, in general the problem is neither a convex nor concave program; this makes its solution more difficult.

We define an *interior edge location* to be any point on the edge (boundary) of the location triangle that is not a corner point (input source or market location). We can then prove the following.

Theorem 4.1

If $z_1^*, z_2^* > 0$, then an interior edge location cannot be optimal for (DL4.1) or (PL4.1).

Proof: We begin with (DL4.1). Let $z_1^*, z_2^* > 0$ be the optimal input levels. (Notice that if either $z_i^* = 0$, then the problem reduces to the linear space problem (DL2.2), in chapter 2. In that case "interior edge" locations could be optimal, such as in example 2.3; however, corner locations would be optimal as well.)

(i) To show that the optimal location cannot be on the interior of the edge connecting (a_1, b_1) with (a_2, b_2); call this edge E_{12}. The origin of a coordinate system is arbitrary, so for simplicity let $(a_1, b_1) = (0, 0)$, $(a_2, b_2) = (a_2, 0)$, $(a_0, b_0) = (a_0, b_0)$ with $a_2 > 0$ and $b_0 > 0$. Suppose (x^*, y^*) is an optimal location on the interior of edge E_{12}. With the given coordinate origin, $\partial L/\partial y = r_0 z_0(y^* - b_0)/d_0$, because $y^* - b_1 = 0$ and $y^* - b_2 = 0$. But $r_0 z_0(y^* - b_0)/d_0 \neq 0$, because y^* cannot equal b_0 (unless the input sources and market location are collinear—i.e., the triangle collapses to a line, in which case the result follows from chapter 2). So (4.5) cannot hold, and (x^*, y^*) cannot be an optimal location.

(ii) To show that the optimal location cannot be on the interior of the edge connecting (a_0, b_0) with (a_1, b_1) or on the edge connecting (a_0, b_0) with (a_2, b_2), these are designated edges E_{01} and E_{02}, respectively. It is sufficient to prove this for E_{01}. Again, without loss of generality, we change the coordinate origin so that $(a_0, b_0) = (0, 0)$, $(a_1, b_1) = (a_1, 0)$, $(a_2, b_2) = (a_2, b_2)$ with $a_1 > 0$ and $b_2 > 0$. Suppose (x^*, y^*) is an interior edge point of E_{01}. Then with the given coordinate origin, $\partial L/\partial y = r_2 z_2^*(y^* - b_2)/d_2 \neq 0$ because $y^* \neq b_2$ unless the input and market locations are collinear. Thus, (4.5) cannot hold and (x^*, y^*) cannot be an interior edge location.

For (PL4.1) let z_0^* be the optimal output level. Then (z_1^*, z_2^*, x^*, y^*) must be optimal for (DL4.1) with $\bar{z}_0 = z_0^*$, and the result is immediate.

□

The exclusion of interior edge locations is theoretically interesting but it appears to be of little computational value in practice. For these problems, even though the optimum will lie within the location triangle, we solve the problem using R^2 as the feasible set; thus, interior edge points would not be boundary points from a computational viewpoint. This approach allows us to use differential methods to find interior optima, which can then be compared with the three corner solutions; the set of interior edge solutions does not need to be checked.[5]

Kusumoto [1984] has also characterized conditions under which we can conclude that the optimum location must be an interior point of the location triangle; these are given in the following.[6]

Theorem 4.2

In (DL4.1) if (a) there is some radial direction away from the output market which decreases total cost, and (b) as one moves away from input source i along edge E_{0i} toward the output market, total cost decreases, then the optimal plant location must be an interior point of the location triangle.

Proof: The proof is complex and is omitted; see Kusumoto [1984, theorem 2].

Conditions (a) and (b) simply establish that none of the corner points are local optima, and therefore cannot be global optima, and so by theorem 4.1, an interior point must be optimal. This characterization is intuitively valid, but it does not appear to aid solution of the problem because both an interior search and a separate check of all three corner points is necessary to verify an optimum.

It should be pointed out that theorems 4.1 and 4.2 do not rely on the transport cost functions being linear in distance transported. As long as the transport cost rates are increasing functions of Euclidean distance transported—i.e., $c_i(x, y)$ can be written as $c_i(d_i)d_i$—these theorems hold. This transport cost property, in fact, will almost always hold.

A General *n*-input Model

Although n-input planar models are more difficult to analyze and provide less unambiguous results than the two-input model, they still have received substantial attention in the literature (see Alonso [1967], Hurter and Wendell [1972], Nijkamp and Paelinck [1973], Goldman [1974], Thisse and Perreur [1977], HMV [1980], EKR [1981], Hurter and Venta [1982], Clapp [1983], Shieh [1983a], Venta and Hurter [1985]). Consequently, we will briefly discuss these problems and some of the major properties and issues that have been considered. For brevity we will only consider the design/location problem.

Letting (a_i, b_i) be the source location of input i, and all other notation be the same as above, the firm's planar design/location problem is

$$\text{minimize} \quad C = r_0 d_0(x, y)\bar{z}_0 + \sum_{i=1}^{n} (p_i + r_i d_i(x, y)) z_i \qquad (4.7)$$

$$\text{(DL4.2)} \quad \text{subject to} \quad \bar{z}_0 = F(z_1, \ldots, z_n)$$

$$z_i \geqslant 0 \qquad i = 1, \ldots, n$$

$$(x, y) \in R^2 \qquad (4.2)$$

As stated earlier, it can be shown that the optimal plant location for (DL4.2) must be in the convex hull of $\{(a_0, b_0), \ldots, (a_n, b_n)\}$; call this set S_c. The boundary of S_c is made up of straight lines connecting the outermost (a_i, b_i) (i.e., S_c is a polytope); these (a_i, b_i) will be called extreme points. *Interior edge points* are now defined as those points on the boundary of S_c which are not extreme points. Let E_{ij} designate the edge connecting extreme points (a_i, b_i) and (a_j, b_j).

Theorem 4.3

If at least two inputs are used at positive levels and their source locations are not all collinear with the output market, then no interior edge point can be an optimal location for (DL4.2).

Proof: Let E_{jk} be an arbitrary edge. As in the proof of theorem 4.2, we adjust the coordinate system so that $(a_j, b_j) = (0, 0)$, $(a_k, b_k) = (a_k, 0)$, and all other (a_i, b_i) have $b_i \geqslant 0$. Then

$$\partial L/\partial y = r_0 z_0 d_0^y + \sum_{i=1}^{n} r_i z_i^* d_i^y < 0$$

because $d_i^y = 0$ for any i with $b_i = 0$, and $d_i^y < 0$ for any i with $b_i > 0$ (and the noncollinearity assumption implies that for some i, $z_i > 0$ and $b_i > 0$). Thus the necessary condition that $\partial L/\partial y = 0$ cannot hold on the interior of E_{jk}. □

Theorem 4.3 is a generalization of the two-input edge exclusion result in theorem 4.1 and it holds also for the case where output is variable. Thus, in general, the optimal location will be either at a corner of S_c or in the interior of S_c. Unlike the triangular case, the optimal location could be at a market or input source location (a_i, b_i) and also be in the interior of S_c.

Alonso [1967] has investigated the production and location interaction and the relative likelihood of some (a_i, b_i) being optimal vis a vis a comparable Steiner-Weber problem. His investigation created some misunderstanding concerning the differences between the locational pulls associated with the Weber problem and those with the design/location problem and the general relationship between the two problems. This misunderstanding was later resolved by Nijkamp and Paelinck [1973], Thisse and Perreur [1977] and Shieh [1983a]. Because this relationship is of conceptual and computational interest, we will briefly consider it now.

The traditional approach to analyzing the optimum for (DL4.2) and similar production/location problems has been to think of the optimization as proceeding in two steps. First, at each location (x, y) the optimal input levels for that location and output level, call them $z_i(x, y; \bar{z}_0)$, and the corresponding total cost, $C(x, y; \bar{z}_0)$, can be derived (at least in theory). (The arguments x, y, and $\bar{z}_0$ will be suppressed unless needed for clarity.) Letting $P_i = p_i + r_i d_i$, we can then find the spatial optimum by differentiating C with respect to the location variables x and y:[7]

$$\partial C/\partial x = r_0 \bar{z}_0 d_0^x + \sum_{i=1}^{n} P_i \frac{\partial z_i}{\partial x} + \sum_{i=1}^{n} r_i z_i d_i^x = 0 \tag{4.8}$$

$$\partial C/\partial y = r_0 \bar{z}_0 d_0^y + \sum_{i=1}^{n} P_i \frac{\partial z_i}{\partial y} + \sum_{i=1}^{n} r_i z_i d_i^y = 0 \tag{4.9}$$

(Because the results with respect to y are identical to those for x, we will do the analysis only in terms of x.) The first term in (4.8) is the "market pull", while the third term is the "material pull." These two terms together represent what is called the "Weberian pull." They represent the economic forces on the plant location due to the volume of materials (output and inputs) transported times the marginal transport costs as location changes. The name comes from the fact that the optimality conditions for the Steiner-Weber problem require that the Weberian pull be zero; i.e.,

$$\partial C/\partial x = r_0 \bar{z}_0 d_0^x + \sum_{i=1}^{n} r_i z_i d_i^x = 0 \tag{4.10}$$

(See note 7.) The second term in (4.8), which is called the "Moses-Predohl pull" [Nijkamp and Paelinck 1973], represents the economic pull on the optimal location caused by the *substitution of inputs* resulting from delivered factor price changes caused by changes in x. As Alonso [1967, p. 30] points out, if no substitution of inputs is allowed, which is the case for (SW), then the Moses-Predohl pull is zero and so (DL4.2) reduces to (SW) and (4.8) reduces to (4.10). Alonso goes on to explain the implications when the Moses-Predohl pull is nonzero, and concludes that the Moses-Predohl pull would equal zero only if all the substitution effects cancel out. He implies that this would be an unusual occurrence and that in general the optimal location for (DL4.2) would not be the transport cost minimizing (i.e., Weberian) location. However, Nijkamp and Paelinck [1973] demonstrate that for an optimum—not at an (a_i, b_i)—to (DL4.2), the Moses-Predohl pull will *always* be zero. This can be seen most easily by following Shieh's [1983a] recommendation to use a one-step partial differentiation of C with respect to x, y, z_i, and λ (which is the approach we have used in this book), rather than the two-step approach normally used. In that case the first-order condition with respect to x for an optimum location becomes (4.10), with each z_i replaced by the optimal input level, z_i^*. The principal conclusion of the discussion of locational pulls is that the optimal plant location for (DL4.2) is identical to the optimal plant location for (SW) *with input levels set at the optimal levels*, z_i^*; thus, as pointed out by Thisse and Perreur [1977], every optimal location for (DL4.2) is an ex post transport cost minimizing location (i.e., an optimum for a corresponding Steiner-Weber problem).

Notice that we have already used this result in the previous chapters. However, to differentiate the Weberian pull, where input levels are set a priori, from the comparable condition for production/location problems, where the optimal input levels are derived within the model, we have called the pull resulting from the latter case the "net locational pull" (NLP). In one sense the NLP is a Weberian pull, but it is quite different because it is a Weberian pull for a Weber problem which cannot be determined until the original production/location problem is solved.

It should be noted that equivalent properties hold for profit maximization versions of (DL4.2). Only minor changes are required to prove comparable results.

Conceptually, extending one-facility, planar production/location models to a multifacility structure is straightforward and comparable to the same generalization for network spaces. Unfortunately, except for a brief discussion in Hurter and Venta [1982], this generalization has not been done, probably because of its computational complexity and the difficulty in obtaining instructive sensitivity analyses. We will formulate a multifacility version of (DL4.2) here because there are, in fact, some readily provable production/location properties for this model and the multifacility version of (SW) has received attention in the literature.

Retaining the notational conventions used in chapter 3, let

K = the number of plants to be located by the firm
(x^k, y^k) = the location of the kth plant
M = the number of spatially distinct customer markets
(a_m, b_m) = location of customer market m
z_0^k = the amount of output produced at plant site k
z_{0m}^k = the amount of output produced at plant site k and sent to customer location m
(a_i, b_i) = the location of input source i
z_i^k = amount of input i used by plant k

If the amount of output being sent to each market m is fixed at some level, D_m, then the firm's problem becomes the planar equivalent of (DL3.3). (As mentioned in chapter 3, the values of the z_0^k's may or may not be fixed a priori; we use the more general form here.)

$$\underset{z_i^k, z_{0m}^k, x^k, y^k}{\text{minimize}} \quad C = \sum_{k=1}^{K} \left[\sum_{m=1}^{M} r_0 d_0(x^k, y^k) z_{0m}^k + \sum_{i=1}^{n} (p_i + r_i d_i(x^k, y^k)) z_i^k \right] \tag{4.11}$$

(DL4.3) subject to

$$z_0^k = \sum_{m=1}^{M} z_{0m}^k = F(z_1^k, \ldots, z_n^k) \quad \text{for } k = 1, \ldots, K$$

$$\sum_{k=1}^{K} z_{0m}^k = D_m \quad \text{for all } m$$

$$z_i^k \geqslant 0 \quad \text{for all } i, k$$

$$(x^k, y^k) \in R^2 \quad \text{for all } k = 1, \ldots, K$$

Theorem 4.4

No interior edge point of S_c can be an optimal plant location for (DL4.3) if the plant uses positive amounts of at least two inputs and not all source locations for these inputs are collinear with the output markets.

Proof: The result follows easily by combining the argument preceding theorem 3.4 with the proof of theorem 4.3. □

This theorem implies that the optimal locations are either at extreme (corner) points of S_c or in the interior of S_c. The computational usefulness of this result is an unanswered empirical question; further comments on this are reserved until later.

Parametric Analysis

Changes in Output

As mentioned previously, it is quite common for firms to perform their design/location planning for several possible output levels. In general, the optimal input mix and plant location can vary with facility size (capacity). However, if special production function properties are satisfied the optimal input mix and location are invariant with output level. These results, and their proofs, are identical to those in chapters 2 and 3, so they will simply be stated here, without proof, for completeness. (The theorems will be stated in terms of (DL4.2) and (DL4.3) and they obviously hold for the special case of (DL4.1).)

Theorem 4.5

In problem (DL4.2) [(DL4.3)], if output transport costs are nill or independent of location and F is homothetic, then the optimal plant location(s) and input ratios are independent of the output level (independent of proportional changes in the output or demand levels).

Theorem 4.6

In problem (DL4.2) [(DL4.3)] if F is homogeneous of degree one, then if $(\mathbf{z}, x, y)$ $[(Z, \mathbf{x}, \mathbf{y})]$ is an optimal solution vector for output level $\bar{z}_0$ [demand vector $(D_1, \ldots, D_M)$], then $(\alpha\mathbf{Z}, x, y)$ $[(\alpha Z, \mathbf{x}, \mathbf{y})]$ is optimal for output level $\alpha\bar{z}_0$ [demand vector $(\alpha D_1, \ldots, \alpha D_M)$].

An alternative approach to proving theorems 4.5 and 4.6 has been developed by Hurter and Venta [Hurter and Venta 1982; Venta and Hurter 1985]. This approach is based upon a special form of problem separability which exists for (DL4.2) when the production function satisfies certain conditions; this separability is sequential rather than absolute (see Hurter and Wendell [1972], Goldman [1974], HMV [1980], Clarke [1984], and HMV [1984] for discussions of this form of separability).

Specifically, Hurter and Wendell [1972] recognized that when $r_0 = 0$ in (DL4.2) the following holds. For any fixed location, $(\bar{x}, \bar{y})$, (DL4.2) reduces to a subproblem in the z_i's only, which has the following first-order conditions for an optimal solution $\mathbf{z}^*$.

$$p_i + r_i d_i(\bar{x}, \bar{y}) - \mu(\bar{x}, \bar{y}) F_i(\mathbf{z}^*) = 0 \qquad i = 1, \ldots, n \tag{4.12}$$
$$\bar{z}_0 = F(\mathbf{z}^*)$$

where $\mu(\bar{x}, \bar{y})$ is the Lagrangian multiplier which measures the change in the optimal objective function value associated with a change in $\bar{z}_0$ at $(\bar{x}, \bar{y})$; it can be interpreted as the marginal cost of production at $(\bar{x}, \bar{y})$.

If we then multiply through each equation in (4.12) by z_i^*, (4.12) can be rewritten as

$$[p_i + r_i d_i(\bar{x}, \bar{y})]z_i^* = z_i^* \mu(\bar{x}, \bar{y}) F_i(\mathbf{z}^*) \qquad i = 1, \ldots, n \tag{4.13}$$

Adding these equations together, for each input, yields for every $(\bar{x}, \bar{y})$

$$\sum_{i=1}^{n} [p_i + r_i d_i(\bar{x}, \bar{y})]z_i^* = \sum_{i=1}^{n} z_i^* \mu(\bar{x}, \bar{y}) F_i(\mathbf{z}^*) \qquad i = 1, \ldots, n \tag{4.14}$$

The left-hand side of (4.14) has the same form as the objective function in (DL4.2) with $r_0 = 0$, except using the optimal input mix at a specified location. This justifies rewriting the objective of (DL4.2) as

$$\underset{(x,y)\in S_c}{\text{minimize}} \qquad \mu(x, y)\left[\sum_{i=1}^{n} z_i^* F_i(\mathbf{z}^*)\right] \tag{4.15}$$

Now homothetic production functions have the property that $\Sigma_{i=1}^{n} z_i F_i(\mathbf{z}) = K$ for *all* $\mathbf{z}$ such that $F(\mathbf{z}) = \bar{z}_0$ where K is some positive constant. It then follows that if F is homothetic, (DL4.2) can be replaced by the problem

$$\underset{(x,y)\in S_c}{\text{minimize}} \qquad \mu(x, y) \tag{4.16}$$

which is a pure location problem, once the $\mu(x, y)$ are known. Notice, that this procedure is really saying that we solve a production problem at every (x, y) in the feasible location space to obtain $\mu(x, y)$ and then choose the location with the smallest $\mu(x, y)$. (This is the same idea as that discussed in chapter 3 for network problems; in that case, however, only a finite number of (x, y)'s need to be considered.) Clarke [1984] has pointed out that this, in theory, can be done implicitly for any production function, subject to certain regularity conditions. The crucial factor (as pointed out by Hurter, Martinich, and Venta [1984], however, is that homotheticity guarantees this regularity; more importantly, this separation is meaningful only if the $\mu(x, y)$ can be written *explicitly* (in closed form) in terms of (x, y), which homotheticity allows. In fact, Hurter and Venta have shown that μ is independent of $\bar{z}_0$, and this separation can be expressed generally using the following.

Lemma

In (DL4.2) let $r_0 = 0$ and define $C(x, y; \bar{z}_0)$ to be the minimum cost of producting $\bar{z}_0$ units of output at location (x, y). If F is homothetic, then $C(x, y; \bar{z}_0)$ can be written as $C(x, y; \bar{z}_0) = h(x, y)g(\bar{z}_0)$, where h depends explicitly only on the location and g depends only on the output rate.

Proof: By definition (see Shephard [1970] or Goldman [1974]), F can be written as $F(\mathbf{z}) = f[H(\mathbf{z})]$ where $H(\mathbf{z})$ is a production function homogeneous of degree one and f is a positive, differentiable, monotonic transformation. Let

$$h(x, y) = \min \sum_{i=1}^{n} [r_i d_i(x, y) + p_i] z_i$$

$$\text{subject to} \qquad H(\mathbf{z}) = 1$$

that is, $h(x, y)$ is the minimum cost at (x, y) of producing one unit of output expressed in terms of H. Notice that because H is degree-one homogeneous the cost of producing an arbitrary rate of output, $\hat{z}_0$, expressed in terms of H, is $\hat{z}_0 h(x, y)$. But $f[H(\mathbf{z})] = \bar{z}_0$, or $\hat{z}_0 = H(\mathbf{z}) = f^{-1}(\bar{z}_0) \equiv g(\bar{z}_0)$. Thus, $C(x, y, \bar{z}_0) = h(x, y)g(\bar{z}_0)$. □

In other words, the cost of producing $\hat{z}_0$ units of output expressed in terms of H is $h(x, y)\hat{z}_0$, and the cost of producing $\bar{z}_0$ units of output expressed in terms of F is $h(x, y)g(\bar{z}_0)$. So $g(\bar{z}_0)$ is the rate of output in terms of H that would cost the same as producing $\bar{z}_0$ using F. Thus, with $r_0 = 0$, (4.7) can be written as

$$\underset{x,\, y}{\text{minimize}} \quad h(x, y)g(\bar{z}_0) = \underset{x,\, y}{\text{minimize}} \quad h(x, y) \tag{4.17}$$

which suggests the sequential separability mentioned earlier. Theorem 4.5 (with respect to (DL4.2)) follows immediately from (4.17) because, (1) the optimal location is independent of output level; and (2), because of (1), nonspatial production theory ensures that the optimal input ratios will be invariant with output level if F is homothetic.

When output transport costs are included, the previous lemma can be used to rewrite (4.7) as

$$\underset{x,\, y}{\text{minimize}} \qquad r_0 d_0(x, y)\bar{z}_0 + h(x, y)g(\bar{z}_0) \tag{4.18}$$

Because $g(\bar{z}_0)$ is not necessarily proportional to $\bar{z}_0$, the location, (x, y), that minimizes (4.18) will not necessarily be the same for all output levels. However, if F is homogeneous of degree one, then $g(\bar{z}_0) = \bar{z}_0$ so that (4.18) can be written as

$$\underset{x,\, y}{\text{minimize}} \qquad [r_0 d_0(x, y) + h(x, y)]\bar{z}_0 \tag{4.19}$$

Notice that $h(x, y)$ is the average cost of production and transportation of the inputs at (x, y) for all levels of output. Theorem 4.6 (for DL4.2) then follows from (4.19) in the same way that theorem 4.5 followed from (4.17).

The same type of separability extends to multiplant and multi-market models, such as variations of (DL4.3). For brevity the details are omitted here (see Hurter and Venta [1982]), but theorems 4.5 and 4.6 can be shown to hold for these cases as well. Separability for this case will be discussed later, however, with respect to possible solution methods for multiplant problems.

It is known from earlier chapters that the invariance properties of theorems 4.5 and 4.6 hold even if the transport costs are nonlinear in distance shipped as long as they are linear in quantity shipped. These theorems also require that prices be constant; when prices are a function of quantity purchased (or sold), these theorems do not hold in general. Hurter and Venta [1982], however, used their separability approach to demonstrate that if the cost of purchasing inputs can be written in the form $p_i(z_i)z_i = a_i + c_i z_i$, where a_i, $c_i > 0$, then these theorems still hold. This pricing mechanism in not unusual in practice; it says that the total cost of purchasing an input is made up of a fixed cost plus a per unit charge. Venta and Hurter [1985] have shown that theorem 4.5 also extends to the case where output price at the market is a function of quantity. Thus, theorems 4.5 and 4.6 are more general than they appear.

If the conditions in theorems 4.5 and 4.6 are not satisfied, then changes in the output level can, but do not necessarily, alter the optimal input mix and plant location. There is a misconception that theorems 4.5 and 4.6 hold only for "interior" optima; in fact, these hold for any optimal location, including extreme points of S_c. In addition, these theorems have been incorrectly stated as "if and only if" results—e.g., see Miller and Jensen [1978] and Mathur [1979]; in fact, locational invariance often holds when an (a_i, b_i) is optimal even when F is not homogeneous of degree one. Martinich and Hurter [1988] prove and illustrate these facts.

Substantial analysis has been performed for (DL4.1) when the production function is homogeneous, but not necessarily of degree one. Our proof of the following result is algebraic and is more direct than the normal differential methods used (see KMB [1974], EKR [1981], and Heaps [1982] for alternative proofs).

Theorem 4.7

In (DL4.1) [and (DL4.2)] let F be homogeneous of degree k. As $\bar{z}_0$ increases the optimal plant location moves "toward" ["away from"] the output market location (a_0, b_0) according as $k > [<]\ 1$.[8]

Proof: We will prove the result for (DL4.1), using the same notation and approach as that used for theorem 2.8; the same proof goes through for (DL4.2). Consider two output levels $z_0^2 > z_0^1$, and let $(\mathbf{z}^j, X^j)$ be an optimal solution for output level z_0^j, where $\mathbf{z}^j = (z_1^j, z_2^j)$ and $X^j = (x^j, y^j)$. Define

$$P^h(X^j) = [p_1 + c_1(X^j)]z_1^h(X^j) + [p_2 + c_2(X^j)]z_2^h(X^j)$$

where $z_i^h(X^j)$ is the optimal value of input i when the plant is located at X^j and z_0 is fixed at z_0^h. (So $P^h(X^j)$ is the cost of purchasing and transporting the optimal amounts of inputs when the plant is located at (x^j, y^j) and output is fixed at z_0^h.) Then, by the optimality of $(\mathbf{z}^1, X^1)$ for $z_0 = z_0^1$,

$$A \equiv [c_0(X^1) - c_0(X^2)]z_0^1 \leqslant [P^1(X^2) - P^1(X^1)] \equiv B \qquad (4.20)$$

Define $t = z_0^2/z_0^1 > 1$. Using the same approach as in theorem 2.8, the optimality of $(\mathbf{z}^2, X^2)$ for z_0^2 implies that

$$A \geqslant t^{[(1/k)-1]}B \qquad (4.21)$$

If $k < 1$, then $t^{[(1/k)-1]} > 1$ and (4.20) and (4.21) can both hold only if $A, B \leqslant 0$. But $A \leqslant 0$ if and only if $c_0(X^1) \leqslant c_0(X^2)$. This implies that X^2 is at least as far from (a_0, b_0) as X^1. So as output increases, the location moves "away" from the output market. If $k > 1$, a similar argument reveals that an increase in output implies that the optimal location must move "toward" the output market. □

Theorem 4.7 is consistent with intuition because as output increases, if $k < 1$, the amount of output per unit of input decreases. This makes the market pull due to transporting the output weaker relative to the input pulls; these changes create a shift away from the output market. When $k > 1$, the opposite forces occur. To illustrate this theorem we present the following example.

Example 4.2

Consider the same location space as that in example 4.1. However, we now let $F(z_1, z_2) = z_1^{0.3} z_2^{0.6}$; $p_1 = p_2 = 20$; $r_1 = r_2 = 1$, and $r_0 = 2$. The optimal solutions for two values of $\bar{z}_0$ are given below.

$\bar{z}_0$	x^*	y^*	z_1^*	z_2^*	C^*
100	0.93	2.43	100.79	214.45	9407.15
150	1.05	3.13	155.23	339.63	14719.72

Notice that the optimal location moves away from $(a_0, b_0) = (0, 0)$ as $\bar{z}_0$ increases. In addition, the movement is *not* along a ray from the output market (i.e., the ratio x^*/y^* changes); the optimal location moves toward the source of input 2 with a corresponding increase in the relative use of input 2. As mentioned above (see note 8), the spatial "movements" in theorem 4.7 do not have to be strict if the optimal location is at some (a_i, b_i). For example, if in example 4.2 we let $r_0 = 3$, then $(x^*, y^*) = (0, 0)$ for $\bar{z}_0 = 50, 100$, and 150 (and z_1^*/z_2^* is constant). Thus, statements of spatial sensitivity must be considered with care if the optimal location is at some (a_i, b_i).

Just as with Theorem 2.8, the conditions on Theorem 4.7 can be relaxed for localized perturbations in output. If F is homothetic then the optimal plant location moves (toward, away from) the output market according as F exhibits locally (increasing, decreasing) returns to scale.

When the production function is not homogeneous (or at least homothetic) then it is not possible to characterize in general the effects of output perturbations on the optimal location (as shown in example 2.5). Theorem 4.7, however, makes it reasonable to conjecture that if output is increased within a range over which F exhibits increasing returns to scale, then we would expect the optimal location for (DL4.1) to move toward the market, and the opposite effect is likely to occur under decreasing returns to scale.

When there are more than two spatially distinct inputs, and the conditions of theorems 4.5 or 4.6 are not satisfied, general statements regarding spatial sensitivity analysis become almost impossible because input markets can surround the output market(s). Although changes in output level may have direct effects on the output market pull similar to those illustrated above, substitution among inputs can now cause spatial effects that partially or totally counteract these forces and can lead to other, seemingly counterintuitive, spatial effects. When dealing with more than two input sources it is usually necessary to use a numerical approach to sensitivity analysis.

Changes in Prices

The production effects of input price changes for planar production/location models have been studied by Heaps [1982]. His primary conclusion concerning changes in input prices is that the inclusion of location in the neoclassical model of the firm does not change the comparative statics results with respect to the optimal input quantities: an increase in the price of input i at its source p_i, decreases the use of input i (and increases the use of input $j \neq i$ if output is fixed). Unfortunately, Heaps does not consider the effect on the optimal plant location which results from such a change. One might conjecture that the optimal plant location will move away from the source of the ith input; this is usually true, but not necessarily, as shown in the following.

Theorem 4.8

Consider (DL4.1). Let $(\mathbf{z}^h, X^h)$ be optimal for prices (p_0, p_1^h, p_2) where $X^h \equiv (x^h, y^h)$ represents the optimal location. If $p_1^2 > p_1^1$, then (i) $z_1^2 \leqslant z_1^1$, (ii) $z_2^2 \geqslant z_2^1$, and (iii) either $d_1(X^2) \geqslant d_1(X^1)$, or $d_1(X^2) \leqslant d_1(X^1)$ and $d_2(X^2) \leqslant d_2(X^1)$. That is, as p_1 increases, either the optimal location moves "away from" (no closer to) the source of input 1 or it moves closer to the sources of both inputs. Equivalent results hold for changes in p_2. (If the optimal locations are not at (a_i, b_i) points, the inequalities in (i) and (ii) are strict.)

Proof: The proofs of (i) and (ii) are identical to those for theorem 2.9, and they have been proved by Heaps [1982]; so they are omitted here for brevity.

(iii) Define $c_i(X^h) = r_i d_i(X^h)$. From (i), let $z_1^2 < z_1^1$ (if $z_1^2 = z_1^1$, the proof of (iii) is trivial). Because X^1 is optimal for prices (p_0, p_1^1, p_2),

$$[c_1(X^1) - c_1(X^2)]z_1^1 \leqslant [c_2(X^2) - c_2(X^1)]z_2^1 + [c_0(X^2) - c_0(X^1)]z_0 \qquad (4.22)$$

Now defining C, D, and E as the quantities in the three brackets, respectively, we can write (4.22) as

$$Cz_1^1 \leqslant Dz_2^1 + Ez_0 \qquad (4.23)$$

Likewise, from the optimality of X^2 for prices (p_0, p_1^2, p_2), we get

$$Cz_1^2 \geqslant Dz_2^2 + Ez_0 \qquad (4.24)$$

Now if $d_1(X^2) \leqslant d_1(X^1)$, then $C \geqslant 0$ and $Cz_1^1 \geqslant Cz_1^2$. So both (4.23) and (4.24) can hold only if $Dz_2^2 \leqslant Dz_2^1$. But $z_2^2 > z_2^1$, which implies $D \leqslant 0$, or equivalently, $d_2(X^2) \leqslant d_2(X^1)$. □

In the two-input model, as the price of one input increases vis a vis the other input the firm substitutes away from that input. The lower use of the input reduces the pull of that input's source location, and the optimal plant location is usually pushed away from that location. This is illustrated by the following.

Example 4.3

Consider example 4.2 with $\bar{z}_0 = 150$, except we will alter the price of the second input. The optimal solutions when $p_2 = 20$ and $p_2 = 25$ (ceteris paribus) are given below.

p_2	x^*	y^*	z_1^*	z_2^*	C^*
20	1.05	3.13	155.23	339.63	14719.72
25	1.07	2.24	178.18	317.00	16356.79

It is possible, according to theorem 4.8, for the substitution of say input 2 for input 1 to result in a move away from the market location which results in the plant being closer to both input markets. An illustration of this can be found in Martinich and Hurter [1988].

The preceding results assumed that the output level was fixed, so changes in output price would have no effect on the input levels and location in these models. If output is a variable as in (PL4.1), a change in the output price *will* normally change the optimal values of the production and location variables. Specifically, if p_0 increases (e.g., due to a demand increase), then the optimal output level will increase (it will not decrease; if a smaller output level is optimal it would have been optimal before the price increase). The effects on z_1^*, z_2^*, and the optimal location are qualitatively the same then as for a parametric increase in output level in (DL4.1). Thus, results equivalent to theorems 4.5–4.7 hold. Specifically, if output transport rates are nill, then a change in the output price does not affect the optimal input ratio and location if F is homothetic. If output transport rates are positive and F is homogeneous of degree $k < 1$—no finite optimum exists if $k \geqslant 1$ in (PL4.1)—then an increase in p_0 causes the optimal plant location to move away from (actually, move no closer to) the output market.

In n-input models, part (i) of theorem 4.8 still holds, but the effects on the other inputs and plant location cannot be characterized in general because a price increase of one input can cause substitution effects among several other inputs. The spatial distribution of the source locations of these inputs can be such that the optimal location moves toward the source location of the input which had a price increase. The movement toward this source location appears anomalous, but it is due to the adjustments of the pulls of the other inputs. The extent to which this phenomenon occurs in practice is not known, but the clear possibility of it occurring suggests that the effects of price changes need to be studied numerically for each individual case.

Changes in Transport Rates

For the most part the comments made in chapter 2 regarding the effects of transport rates in line segment models apply to planar models as well. Both the absolute level of transport rates (i.e., marginal transport costs with respect to distance) and the structure of the transport cost functions are influential in determining the optimal plant design, size, and location. We begin by considering changes in the level of transport rates—i.e., the values of the r_i's in our models.

If the value of r_i is increased for some input i, the delivered price of that input at any *fixed* location increases. However, unlike source price changes, the firm can partially control the delivered price of an input by changing its plant location. For example, suppose r_1 is increased in (DL4.1). Even though the "initial" tendency is to substitute input 2 for input 1, thereby creating additional pull away from the source of input 1, the increase in r_1 affects the net locational pull (NLP) directly:

$$\partial C/\partial j = \sum_{i=0}^{2} r_i d_i^j z_i^* \text{ for } j = x, y$$

This direct effect often forces the optimal location toward the source of input 1, which creates at least a partial substitution back toward input 1. Whether or not the actual use of input 1 increases or decreases depends on the specific situation.

Example 4.4

Consider example 4.2 with $\bar{z}_0 = 150$ and let the value of r_1 vary. The optimal solutions for $r_1 = 1$ and $r_1 = 2$ are given below.

r_1	x^*	y^*	z_1^*	z_2^*	C^*
1	1.05	3.13	155.23	339.63	14719.72
2	1.85	3.15	132.76	367.25	16022.03

In example 4.4 the increase in r_1 causes the optimal location to move toward the source of input 1 (the point (10, 0)), but unlike example 2.6 the amount of input 1 used decreases and the amount of input 2 increases. The first effect (change in plant location) is due to forces acting on the NLP, while the relative use of the inputs is determined by their relative delivered prices (which partially affects the NLP through the z_i^*'s). Although we cannot determine the effects on the optimal plant location and input quantities resulting from a change in r_i, Heaps [1982] has proved that $d_i(x^*, y^*)z_i^*$ is decreasing in r_i. Thus, if z_i^* increases with an increase in r_i, then the plant location will have to move closer to the source of input i.

Changes in the transport rate of the output, r_0, produce more determinate effects. EKR [1981] and Heaps [1982] have shown that as r_0 increases, the optimal plant location for (DL4.1) cannot move farther from the output market. With input transport rates, input substitution can be used to reduce the effect of transport rate increases; however, the firm can reduce the effect of output rate increases in design/location problems only by moving closer to the output market.

In addition to the transport rate levels, the actual structure of the transport cost functions are important. Throughout this chapter we have assumed that the transport cost functions for input (or output) i were of the form $c_i(d_i(x, y), z_i) = r_i d_i(x, y) z_i$; that is, transport costs are linear in both distance and quantity shipped. We used this assumption because of its almost universal use in the research literature and because the Steiner-Weber problem has the same assumed transport cost structure. More importantly, this assumption greatly simplifies the analysis and makes it easier to gain insight into the production/location interactions. Our conjecture is that this insight is generalizable to a reasonable extent. However, this assumption is by no means irrelevant and, in fact, holds infrequently in practice. Therefore it is of value to examine the extent to which some of the earlier results might be affected by relaxing this assumption.

First of all, the concavity (including linearity) of the transport cost function in distance shipped was an important property for proving the endpoint and node optimality results in chapters 2 and 3. The interior edge exclusion theorems in this chapter do not require such restrictions. For example, if all of the transport rates, r_i, are increasing in distance shipped, the proofs hold with minor changes, and interior edge points cannot be optimal plant locations for the models presented in this chapter. (In fact, under relatively general conditions, these exclusion results hold.) The concavity or convexity of the transport cost functions do, however, affect the likelihood of interior points of S_c being optimal. As demonstrated in chapter 2, convexity of transport costs in distance tends to promote interior optimal locations, whereas concavity in distance tends to promote extreme-point optima. However, it appears that convexity or concavity properties of individual transport cost functions cannot alone guarantee interior or extreme-point optimality in general; other conditions are needed. In addition, the discussions in the preceding parts of this section do not necessarily rely upon the linearity of transport costs in distance.

The comments made concerning parametric changes are usually valid as long as the transport costs are increasing in distance (see Martinich and Hurter [1988]).

In contrast to nonlinearities in distance, nonlinearity of transport costs in quantity is crucial for at least one set of results. Miller and Jensen [1978], Shieh and Mai [1984], and Ziegler [1986] have shown that if transport costs are not linear in quantity shipped (e.g., quantity discounts), then the invariance results for changes in output do not hold. (An interesting example of transport cost nonlinearity is in energy facilities, where transmission line losses create nonlinearities [Clarke and Shrestha 1983] and Shieh [1985].) This fact has already been clearly illustrated in example 2.7. For the multifacility case Hurter and Venta [1982] showed that invariance holds when transport costs are nonlinear in quantity only under extremely special conditions.

Economies and diseconomies of scale in transport costs can also reinforce or counteract the effects due to production economies or diseconomies of scale. For example, in theorem 4.7 diseconomies of scale in production push the optimal plant away from the output market location as output level increases. If there are economies of scale in transporting the output, however, this movement will be reduced and it is possible that the diseconomies in production could be totally overcome so that the optimal plant location could move toward the output market as output increases. Economies and diseconomies of scale in transport can also reinforce or counteract the effects of other parametric perturbations. Nonlinearity of transport costs in quantity affects the earlier parametric results, but the effects can often be estimated in advance under specific conditions. The unfortunate conclusion, however, is that parametric analysis often has to be performed on a case-by-case basis and that broadly general results are not achievable.

Solution Methods

In this section we present three approaches to solving planar production/location problems. The first is a practical, yet cumbersome grid search; the second uses a separation of the production and location problems to convert the problem to a Steiner-Weber type form; the third, which is most valuable for multifacility problems, is an "alternating" heuristic method similar to that suggested in chapter 3, which solves Steiner-Weber problems as subproblems. Because all of the following methods are either based upon solution of Steiner-Weber problems or have similar features, we begin with a brief discussion of the solution methods for solving Steiner-Weber problems.

Solving Steiner-Weber Problems

The Steiner-Weber problem (SW) was formulated in chapter 1. It is identical to (DL4.2) except that the z_i's are fixed in value a priori, so that it is a pure location problem in x and y only. The natural approach to solving this problem is to compute the partial derivatives of the objective function (1.10) with respect to x and y and set them equal to zero. Assuming that $(x, y) \neq (a_i, b_i)$ for some i, the partial derivatives are

$$\partial C_{SW}/\partial x = \sum_{i=0}^{n} r_i z_i d_i^x = 0 \tag{4.25}$$

$$\partial C_{SW}/\partial y = \sum_{i=0}^{n} r_i z_i d_i^y = 0 \tag{4.26}$$

Remember that $d_i^x = (x - a_i)/[(x - a_i)^2 + (y - b_i)^2]^{1/2}$ and $d_i^y = (y - b_i)/[(x - a_i)^2 + (y - b_i)^2]^{1/2}$, so that (4.25) and (4.26) are undefined when $(x, y) = (a_i, b_i)$. Because of the convexity of (1.10), (4.25) and (4.26) are necessary and sufficient conditions for an optimum, *if the optimum is not at some* (a_i, b_i). Unfortunately, there is no guarantee that the optimum will not be at an (a_i, b_i). A modification of this approach (due to Weiszfeld [1937], Kuhn and Kuenne [1962], and Cooper [1967]) has been developed which uses the following "modified" partial derivatives.

$$R(x, y) = (\partial C(x, y)/\partial x, \partial C(x, y)/\partial y) \qquad \text{if } (x, y) \neq (a_i, b_i)$$

and if $(x, y) = (a_k, b_k)$, then

$$R(x, y) = \begin{cases} [((u_k - w_k)/u_k)s_k, ((u_k - w_k)/u_k)t_k] & \text{if } u_k > w_k \\ (0, 0) & \text{if } u_k \leqslant w_k \end{cases}$$

where

$$s_k = \sum_{\substack{i=0 \\ i \neq k}}^{n} w_i(a_k - a_i)/[(a_i - a_k)^2 + (b_i - b_k)^2]^{1/2}$$

$$t_k = \sum_{\substack{i=0 \\ i \neq k}}^{n} w_i(b_k - b_i)/[(a_i - a_k)^2 + (b_i - b_k)^2]^{1/2}$$

$$u_k = (s_k^2 + t_k^2)^{1/2}$$

$$w_i = r_i z_i$$

The s_k's and t_k's are simply the partial derivatives of C with respect to x and y with the effects of the kth source deleted (i.e., we eliminate the undefined term in the partials) and the u_k's are a kind of averaging. Thus, the u_k's measure the net pull of all the sources other than source k. So if $u_k > w_k$, then the pull of source k is not sufficient to establish the optimum location at (a_k, b_k); if $u_k \leqslant w_k$, the optimum is at (a_k, b_k). $R(x, y)$ is defined everywhere in S_c and Kuhn and Kuenne [1962] have proved that a necessary and sufficient condition for (x^*, y^*) to be optimal is that $R(x^*, y^*) = (0, 0)$. Thus, some (a_k, b_k) is optimal if and only if $u_k \leqslant w_k$.

One approach to finding the optimum, therefore, is to first compute $R(x, y)$ at each (a_i, b_i); if one of them is zero we can stop—otherwise we can use a standard differential algorithm to find the optimum. This is unnecessarily time consuming. Instead, we can use an interative approach based upon the modified gradient $R(x, y)$ in the following way: Defining $g_i(x, y) = w_i/d_i(x, y)$, it follows from (4.25) that

$$x^* = \left[\sum_{i=0}^{n} a_i g_i(x^*, y^*)\right] \Big/ \left[\sum_{i=0}^{n} g_i(x^*, y^*)\right] \tag{4.27}$$

Similarly, from (4.26),

$$y^* = \left[\sum_{i=0}^{n} b_i g_i(x^*, y^*)\right] \Big/ \left[\sum_{i=0}^{n} g_i(x^*, y^*)\right] \tag{4.28}$$

As long as $g_i(x, y)$ is defined these conditions can be used interatively as follows:

$$x^k = \left[\sum_{i=0}^{n} a_i g_i(x^{k-1}, y^{k-1})\right] \Big/ \left[\sum_{i=0}^{n} g_i(x^{k-1}, y^{k-1})\right] \tag{4.29}$$

$$y^k = \left[\sum_{i=0}^{n} b_i g_i(x^{k-1}, y^{k-1})\right] \Big/ \left[\sum_{i=0}^{n} g_i(x^{k-1}, y^{k-1})\right] \tag{4.30}$$

Of course, g_i is not defined at the (a_i, b_i). However, if we redefine g_i as $g_i(x, y) = w_i/(d_i(x, y) + \varepsilon)$ where ε is a small positive number, then g_i is defined everywhere and as ε approaches zero the new form approaches the original form. (This "ε-perturbed" method is called the hyperboloid approximation method; see Francis and White [1974].) Then, starting with some (x^0, y^0) (4.29) and (4.30) can be used to generate new points, (x^1, y^1), $(x^2, y^2), \ldots$ The procedure continues until no significant improvement in total cost occurs or until Kuhn and Kuenne's optimality conditions are satisfied. This approach has been shown to converge to the optimum and does so very efficiently (see Katz [1969]).

We now illustrate the algorithm with a numerical example.

Example 4.5

Consider the location space in examples 4.1–4.4, but *set* $z_1 = 155$, $z_2 = 340$, and $\bar{z}_0 = 150$, and let $r_0 = 2$, $r_1 = r_2 = 1$. Therefore, in the notation of the algorithm, $w_0 = 300$, $w_1 = 155$, $w_2 = 340$. We will use $\varepsilon = 0.005$ as our perturbation factor in (4.29) and (4.30).

Using standard practice we choose our starting location, (x^0, y^0) to be the center of gravity point:

$$x^0 = \left[\sum_{i=0}^{2} w_i x_i\right] \Big/ \left[\sum_{i=0}^{2} w_i\right]$$
$$= 1.73$$

Similarly, $y^0 = 3.80$. We now compute $g_0(1.73, 3.80) = 300/4.180$, $g_1(1.73, 3.80) = 155/9.106$ and $g_2(1.73, 3.80) = 340/6.441$. Using (4.29) and (4.30) these then generate $(x^1, y^1) = (1.20, 3.73)$. Repeating this again gives the following:

k	(x^k, y^k)	$C(x, y)$
0	(1.73, 3.80)	4851.40
1	(1.20, 3.73)	4827.45
2	(1.11, 3.65)	4825.69
3	(1.09, 3.58)	4824.99
4	(1.09, 3.52)	4824.54

The optimal solution is actually (1.04, 3.13) with $C^* = 4822.86$. (Note that the Weberian pulls, (4.25) and (4.26), approximately equal zero.) If the optimal solution is at some (a_k, b_k), the locations generated by the algorithm will converge to that point. Once convergence becomes apparent, it is reasonable to check explicitly the (a_k, b_k) for optimality using Kuhn and Kuenne's conditions.

In the past 20 years this and other algorithms have been generalized to other distance measures—including mixed distance measures—and to multiple facilities (e.g., see Cooper [1967], Francis and Cabot [1972], Eyster, White, and Wierwille [1973], Planchart and Hurter [1975], Morris and Verdini [1979], Morris [1981], Wesolowsky and Love [1972]).

Grid Search

Because production/location problems do not necessarily have globally concave or convex objective functions—as does (SW), classical methods (e.g., steepest descent) that identify local optima do not necessarily guarantee global optimality. One way around this is to determine the optimal production (design) solution at points distributed over the feasible location set. From this information it is usually possible to determine where local optima exist, and then more refined search around the areas of the local optima can be performed to find a global optimum.

Nijkamp and Paelinck [1973] propose such a numerical search procedure for (DL4.1), which is easily extendable to other one-facility problems, such as (PL4.1) and (DL4.2). Their approach uses the following facts. (1) Any optimal location that is not at some (a_i, b_i) must satisfy the first-order necessary conditions (4.4) and (4.5) [(4.8] and (4.9) for (DL4.2)]; that is the NLP in each coordinate direction must be zero. (2) The points (a_i, b_i) can be optimal, in which case the NLP's do not exist (they can be defined in the form of a subgradient, in which case they need not equal zero); the search must include these points as possible optima. Their algorithm is given below in terms of solving (DL4.1); it is straightforward to adapt this to other one-facility planar models.

Nijkamp and Paelinck Grid-Search Algorithm

1. A "coarse" rectangular grid is constructed on the feasible set.
2. At each intersection point (x, y) on the grid (we will call these *grid points*) the production (design) problem is solved using the delivered prices at that point. The resulting total cost, $C(x, y)$ and the locational pulls (l.h.s. of (4.4) and (4.5)), call these $LP^x(x, y)$ and $LP^y(x, y)$, are computed at each grid point.
3. The set of total costs are inspected to identify subareas with relatively low total costs. On each subarea a "finer" grid is constructed (grid points are made closer) and step 2 is repeated. (The *LP*'s are used to locate the centers of the subarea grids.)
4. Steps 2 and 3 are repeated until further refinement of the grid mesh size does not produce significant reduction in total cost.

Notice that this procedure includes the source and market locations in the search process, and so they are not ignored as with a pure differential method. The $LP(x, y)$ values are used in two ways: (1) to verify whether an "interior" point other than an (a_i, b_i) is optimal, in which case the *LP*'s should approximately equal zero, and (2) to guide the selection and construction of finer grids. For example, if at some location (x, y) $LP^x(x, y) > 0$ and $LP^y(x, y) < 0$, then a local optimum around (x, y) is likely to be in direction with smaller value of x and larger y; so the grid might be centered at some point $(x - \delta, y + \delta)$ where $\delta > 0$, rather than at (x, y) itself.

There are two major drawbacks to this type of approach: (1) it can be computationally time consuming if there are several local optima and each production problem is difficult to solve, and (2) if the original grid is too "coarse," the optimal location can be inadvertently missed in the refined search because the intersection points around the optimum might have much higher total costs than the optimum. Both drawbacks can be controlled by careful grid construction and analysis. From our experience, the objective functions are relatively flat around local optima so that unless one tries to "get away with" an excessively large grid mesh size initially, the global optimum will not be missed. The locational pull values can be especially helpful in pointing toward hidden optima.

We will now illustrate the algorithm by showing how it would solve the problem in example 4.4 with $r_1 = 2$.

1. We construct a grid over the location triangle with a mesh size of 0.5; i.e., points $(0, 0)$, $(0, 0.5)$, $(0, 1)$, ..., $(0.5, 0)$, $(0.5, 0.5)$, ..., $(9.5, 0.5)$, $(10, 0)$ are grid points.

2. At each grid point we compute the production solution, total cost, and locational pulls. The values for a sample of points (most are around the optimum) are given below.

(x, y)	z_1	z_2	$C(x, y)$	$LP^x(x, y)$	$LP^y(x, y)$
$(0, 0)$	136.05	362.79	16325.55	N/A	N/A
$(0, 0.5)$	134.47	364.90	16297.02	−268.59	−52.04
$(1.5, 3.0)$	131.68	368.76	16028.77	−36.94	−4.63
$(1.5, 3.5)$	129.29	372.15	16030.71	−37.26	11.55
$(2.0, 3.0)$	134.29	365.15	16023.69	15.21	−7.21
$(2.0, 3.5)$	131.83	368.55	16025.59	15.66	13.87
$(0, 10.0)$	91.58	442.20	16265.98	N/A	N/A
$(10.0, 0)$	235.39	275.78	17123.65	N/A	N/A

An inspection of (all) the grid-point results indicates the total cost function is convex and has a global optimum in the general range of $1.5 < x < 2.0$ and $3.0 < y < 3.5$. The locational pulls at $(1.5, 3.0)$ are -36.94 in x and -4.63 in y, which suggests $x^* > 1.5$ and $y^* > 3.0$. Likewise, at $(2.0, 3.5)$ the locational pulls are 15.66 in x and 13.87 in y, which suggests $x^* < 2$ and $y^* < 3.5$. These are reinforced by the locational pulls by other surrounding grid points.

3. We then construct a more refined grid with a mesh size of 0.1 over the square $1.5 \leqslant x \leqslant 2.0$, $3.0 \leqslant y \leqslant 3.5$. The values for some of the grid points are given below.

(x, y)	z_1	z_2	$C(x, y)$	$LP^x(x, y)$	$LP^y(x, y)$
(1.8, 3.1)	132.74	367.28	16022.21	−5.01	−2.10
(1.9, 3.1)	133.27	366.55	16022.21	5.12	−2.38
(1.8, 3.2)	132.26	367.95	16022.21	−5.21	1.91
(1.9, 3.2)	132.78	367.22	16022.19	4.96	1.83

4. We then construct a new grid over the range $1.8 \leqslant x \leqslant 1.9$, $3.1 \leqslant y \leqslant 3.2$ using a mesh size of 0.01. The optimum solution (to within 0.01 in either direction) is then

(x, y)	z_1	z_2	$C(x, y)$	$LP^x(x, y)$	$LP^y(x, y)$
(1.85, 3.15)	132.76	367.25	16022.03	−0.02	−0.16

Notice that the NLP's for the optimum are approximately zero, thereby satisfying the first-order necessary conditions (4.4) and (4.5). The second-order conditions have been verified numerically using the grid analysis.

As computers have become faster and capable of handling more information, this approach has become more and more tractable in a practical sense. The major limitation to its effectiveness now is the actual solution of the production subproblems. If the number of inputs is relatively large the solution of these problems can be a difficult problem in itself, although they usually have the advantage of being convex programs that can be solved numerically with standard nonlinear programming methods.

Sequential Separability and an Iterative Algorithm

It has already been mentioned repeatedly that the optimal solutions for many design/location problems exhibit an invariance to changes in output when the production function(s) satisfies certain conditions: usually either homotheticity or degree-one homogeneity. It has been shown earlier that under these same conditions, a natural form of "sequential separability" could be established which allows one to solve the problem by first converting it to a pure location problem and then solving the location problem. Although the first step of converting the problem to a pure location problem is nontrivial, Hurter and Wendell [1972] (also see Nijkamp and Paelink [1973]] performed this separation for the Cobb-Douglas production function:

$$F(z_1, \ldots, z_n) = C z_1^{a1} \ldots z_n^{an}$$

with $C > 0$ and $a1, \ldots, an \geqslant 0$. They showed that $\mu(x, y)$ in (4.16) reduces to

$$\mu(x, y) = 1/z_0\left\{(z_0/C)\left[\prod_{i=1}^{n} |(p_i + r_i d_i(x, y))/ai|^{ai}\right]\right\}^{(\Sigma ai)^{-1}} \tag{4.31}$$

Although minimizing (4.31) appears to be a messy problem, it actually is equivalent to a simple geometric program for which solution methods are well established (see Duffin, Peterson, and Zener [1967] and Wendell and Peterson [1984]).

Goldman [1974] subsequently performed this separation for CES and some two-stage production functions. He showed that when F is Cobb-Douglas or CES, the "separated" form of (DL4.2) can be written as

$$\text{minimize} \qquad C(x, y) = \sum_{i=1}^{n} C_i(d_i(x, y)) \tag{4.32}$$

where each C_i is a function twice differentiable on S_c except at (a_i, b_i). This has the same form as the Steiner-Weber problem and, therefore, can be solved using the same hyperboloid approximation method, except using the gradient of C rather than C_{SW}. Goldman proves that this method is guaranteed to converge to a local optimum when (4.32) holds. Although Goldman has shown that this approach will converge, there is no reported information on how computationally effective this approach is relative to a grid search, for example. In addition, when output transport costs are nonzero, Hurter and Venta [1982] and Venta and Hurter [1985] have shown that a form of "sequential" separability occurs with a degree-one homogeneous production function, but the consequences on the Goldman algorithm have not as yet been studied.

Despite the lack of computational evidence concerning the Goldman algorithm, it suggests an approach essentially identical to that used for the Weber problem, which can be used *without* special conditions on F and without any explicit a priori representation of the production solutions for each location. Specifically, what we can do is perform the Weiszfeld algorithm with $w_i = r_i z_i$ replaced by $w_i = r_i z_i(x, y)$, where $z_i(x, y)$ is the optimal value of z_i at (x, y). So at every iteration, k, we need to solve a production subproblem at (x^k, y^k) to obtain the corresponding $z_i(x^k, y^k)$. To demonstrate this process, we resolve the problem in example 4.4 with $r_1 = 2$, using this approach.

Because we have no initial z_i's with which to find a center-of-mass point, suppose we begin at the "equally weighted" center-of-mass point; so let $(x^0, y^0) = (3.33, 3.33)$. We then solve (DL4.2) at this location and obtain $z_1(3.33, 3.33) = 140.42$, $z_2(3.33, 3.33) = 357.10$, and $C(3.33, 3.33) = 16118.90$. We then use $z_1(3.33, 3.33)$ and $z_2(3.33, 3.33)$ in (4.29) and (4.30) to obtain the next location. The results are tabulated below.

k	(x, y)	$z_1(x, y)$	$z_2(x, y)$	$C(x, y)$
0	(3.33, 3.33)	140.42	357.10	16118.90
1	(2.52, 3.21)	136.16	362.64	16043.29
2	(2.13, 3.19)	134.07	365.46	16025.89
3	(1.96, 3.18)	133.20	366.65	16022.63
4	(1.89, 3.17)	132.88	367.09	16022.11
5	(1.87, 3.17)	132.77	367.24	16022.05
6	(1.86, 3.17)	132.72	367.31	16022.03

The optimal solution is $(x^*, y^*) = (1.85, 3.15)$, $z_1^* = 132.76$, $z_2^* = 367.25$, $C^* = 16022.03$. Because the production function in this problem is Cobb-Douglas, we know from Goldman [1974] that the algorithm will converge to a local optimum (in this case it is globally optimal as well). The extent to which convergence is assured for general production functions is an open question, but out conjecture is that it is locally convergent—for the following reasons. Notice that at each iteration k, the next location generated is the same location as would be generated by the Weiszfeld algorithm for (SW) using $w_i = r_i z_i(x^k, y^k)$. For (x^{k+1}, y^{k+1}) the z_i^{k+1}'s are generated by solving a cost minimization subproblem, so the total cost decreases. It therefore appears to satisfy the conditions of a descent algorithm, and it seems likely that because the Weiszfeld algorithm is locally convergent, the adaptation for (DL4.2) is locally convergent also. We have, in fact, solved several problems using this approach and they always converged to a local optimum.

Even if local convergence can be proved, one problem with this approach is that the local optimum obtained is not necessarily globally optimal, unlike for (SW) where the objective function is convex. This is because design/location problems are not, in general, convex programs (see the discussion preceding theorem 4.1). Consequently, the best use of this approach might be in conjunction with the grid search. Specifically, a coarse grid might first be constructed and analyzed to identify the number and location of possible local optima. Then instead of repeating the grid approach for finer grid sizes, the preceding algorithm could be applied in each region where a local optimum is suspected. The relative effectiveness of this approach versus a pure-grid-search search is an open empirical question.

It is worth reiterating the difference between the Hurter-Wendell-Goldman sequential separability and the adapted Weiszfeld algorithm given above. In the former case (DL4.2) can be rewritten explicitly as a problem in only the location variables. The problem can then be solved directly as a problem in only the location variables. This can be done using the Goldman algorithm or some other method (e.g., geometric programming). The algorithm presented above does *not* rewrite the problem in terms of the location variables. It uses the Weiszfeld iterations to get successive locations, and revises the w_i's at each iteration by solving a production subproblem. Although the Hurter-Wendell-Goldman representation of the problem is conceptually appealing, it is unclear to what extent it provides computational benefit (other than motivating the preceding algorithm).

Alternating Heuristic

For one-facility production/location problems, the grid search and the adapted Weiszfeld algorithm may be effective and computationally tractable in most cases. For multifacilty problems the grid search is essentially undefined, because K facilities must be located simultaneously. At each plant location optimality conditions of the form (4.25) and (4.26) must hold if the location is not at some (a_i, b_i). Thus, an iterative procedure based upon equations of the form (4.29) and (4.30) could be devised for each plant. The major difficulty is that for the multifacility problem there is an embedded allocation problem as well—to determine which customers are served by which plants. If the allocation is set a priori then the problem reduces to a set of K one-facility problems. Unfortunately, this is often not the case, and the allocation problem must be solved simultaneously. Because there has been no published work on this case for production/location problems we can only speculate on a possible approach based upon the work of Cooper [1963, 1964, 1967] for multifacility Steiner-Weber problems.

The general idea is the following. The customers are divided into K groups and assigned to facilities. For each group of customers a one-facility production/location problem can then be solved using one of the preceding methods. Using the new plant locations generated, customers are reassigned to the closest plant. Using this new allocation of customers to plants, K one-facility problems are solved again. This process is repeated until the allocations on two successive iterations are identical. This procedure, called an "alternating" heuristic, does not guarantee achievement of a global, or even local, optimum. However, it does have the property that it is monotonically nonincreasing, and it makes the problem computationally tractable in many cases (see Wendell and Hurter [1976] and Wendell and Rosenblum [1980] for the properties of alternating heuristics). The choice of a starting allocation is certainly crucial in this process and Cooper [1967] makes some suggestions. We have not performed any empirical investigation of this method; the interested reader may find this to be a fruitful area of investigation.

Summary and Conclusion

The generalization of the production/location models in chapter 2 to a planar space is conceptually straightforward and, in many ways, is more interesting than the parallel development for networks in chapter 3. The effects of output transport costs can be isolated and separated from those of the inputs so that the spatial effects of changes in output level, such as those in theorem 4.7, can be characterized. In addition, for the triangular space, the effects of changes in prices and transport rates could be identified and illustrated so as to highlight the interplay between inputs, outputs, and plant location.

For general planar models, however, the effects of parametric changes are generally indeterminable except on a case-by-case basis. Solution of planar production/location problems is complex, but for one-facility problems the grid search and modified Weiszfeld iterative algorithms appear to be very efficient and reliable. Because of this it is possible to perform numerical sensitivity analysis relatively easily. Even though general parametric conclusions cannot be established for many planar models, the numerical sensitivity analysis does provide useful insight. Especially in contrast to the network models in chapter 3, the spatial effects of parametric changes, such as in prices, transport rates, and output, are relatively continuous and informative. Through the study of numerical examples the interplay between the production variables and the location variables is more clearly demonstrated.

Unlike the line segment models, which have been studied extensively (even excessively), little work has been done on planar production/location problems. Because there exists a well-developed body of work for problems of the Steiner-Weber type, it would appear that planar production/location problems might be a promising area of research, especially empirical investigation of solution methods. Unfortunately, up to this point in time, no such study has been published.

Notes

1. There has been a substantial body of research on planar location problems using the "Manhatten" or rectilinear distance measure (see Francis and White [1974]). To date, there has been no published research using rectilinear distances for production/location problems.

2. The convex hull of a set is the smallest convex set containing that set. In this case it is the set enclosed by the lines connecting the input source locations and the market location; this set is referred to as the *location triangle*.

3. KMB [1974], Miller and Jensen [1978] and Shieh and Mai [1984] indirectly consider the possibility of boundary locations being optimal.

4. Except at a corner of the location triangle, where the d_i^x and d_i^y can be undefined, the partial derivatives $\partial L/\partial x$ and $\partial L/\partial y$ must *equal* zero, not just be greater than or equal to zero. This is because the set of feasible locations is all of R^2, not just the location triangle; the fact that the optimum location will lie in the location triangle is a property of the optimum, but not an explicit constraint on the feasible set. At corner points the partial derivatives d_i^* and d_i^y do not exist so that (4.4) and (4.5) will not hold if these points are optima.

5. It should be pointed out that KMB [1974, footnote 1] incorrectly state that the input source locations cannot be optimal plant locations. Their error seems to occur because they use properties regarding the directional derivative of the cost function at these two locations in a radial direction with the output market as the center of the radius. However, this directional derivative does not exist within the location triangle at these points.

6. Kusumoto uses a model with three inputs, where one of the input sources is at the output market; however, his analysis and results are identical to the two-input case presented here.

7. These conditions must hold if the optimal location is not at some (a_i, b_i), in which case the partial derivatives d_i^x and d_i^y will not exist.

8. By "toward" we actually mean "no farther from," and by "away from" we actually mean "no closer to"—i.e., a weak spatial relationship. When the optimal plant location(s) is not at an (a_i, b_i) the movement "toward" and "away from" the output market is a strict movement; the distance to the output market strictly decreases or increases, respectively.

5 STOCHASTIC PRODUCTION/LOCATION MODELS ON A LINE

Because of the extensive investment in plant and equipment, the impacts of plant design and location decisions extend over a period of time. Consequently, many parameters in the problem are not known with certainty at the time the decisions must be made; they must be treated as being uncertain or random. In addition, some of the productive inputs are actually "stock" inputs (e.g., buildings and some equipment) which provide services over time rather than "flow" inputs (e.g., fuel, raw materials). This situation calls for a production/investment/location model that incorporates the uncertainty inevitable in decisions over time. Because of its complexity, however, a complete dynamic model of production/investment/location decisions under uncertainty has not been developed or analyzed, as yet, in the research literature. In this chapter we take the next step toward this general model by incorporating parametric uncertainty into production/location models on a line.

When the construction of a new plant is being considered, input prices, transportation rates, the quantity demanded, and the price of output may not be known with certainty. Furthermore, the technology being designed into the new plant may very well produce a level of output, for given input levels, that is not known with certainty. In short, the production/location problem must be analyzed in an environment in which one or more parameters are random variables. Incorporating uncertainty into production/

location models presents several questions: how should the uncertainty be represented in the model; how should the firm's preferences toward risk be represented; at what point in the decision process do the random parameters become known (realized)? More importantly, what value is there in complicating the model with uncertainty? What types of questions concerning the firm's behavior can be meaningfully investigated? Fortunately, we can draw upon a vast literature on uncertainty in nonspatial economic models of the firm to guide our investigation.

To introduce the reader to models under uncertainty, we present first a brief discussion of von Neumann-Morgenstern (vNM) utility functions. Although not universally accepted as the best way to represent firms' preferences toward risks (gambles), vNM utility functions have a long history of use in economic models under uncertainty. Next some of the relevant nonspatial literature is briefly reviewed. The methods for incorporating uncertainty and firms' risk preferences into production/location models and their subsequent analysis are based upon these nonspatial models. Endpoint optimality results are then developed for stochastic production/location models that are almost identical to those for the deterministic counterparts. This is followed by parametric analyses for design/location and production/location problems, respectively. The additional mathematical complexity created by parametric uncertainty generally leads to weaker results than were derived for deterministic models, but surprisingly strong results can be derived for perturbations in the stochastic aspects of the model: the variability in uncertainty and the firm's aversion toward risk. We also use numerical examples to illustrate how parametric uncertainty and risk preferences can cause nonlinearities in the firm's expansion path and create biases in the resulting cost functions. Brief comments on problem solution and directions for research conclude the chapter.

The incorporation of parametric uncertainty and general risk preferences further complicates the mathematical difficulties inherent in production/location models. Only recently have such models been considered. (Mai [1981] was the first to include uncertainty and Martinich and Hurter [1982] were the first to include general risk preferences with uncertainty.) Work thus far has generally been restricted to location on a line segment or, to a lesser extent, a network. Consequently, the variety of models considered and the extent of analysis in this chapter and in chapter 6 are more limited than those in chapters 2–4. This chapter is intended to review and explain the work that has been published to date; numerous gaps and open questions remain and are under study by researchers, including the authors of this volume.

Parametric Uncertainty and Risk Preferences

Regardless of how parametric uncertainty is introduced into production/location models the preferences of the firm toward risks must be explicitly incorporated in the analysis. The only approach used in the production/location literature (and used widely, though not exclusively, in the non-spatial literature) is to represent firms' preferences using von Neumann-Morgenstern utility functions. These functions have a long history, dating back over two centuries to Daniel Bernoulli. Their full axiomatic development, however, is attributed to von Neumann and Morgenstern [1947], with other major contributions made by Friedman and Savage [1948], Savage [1954], and Luce and Raiffa [1957]. But their most extensive study and use has occurred primarily since 1960 (see Pratt [1964], Arrow [1965] for seminal works and Raiffa [1968], Horowitz [1970] or Keeney and Raiffa [1976] for more extensive introductions and histories).

The principal feature of vNM utility functions is that they convert probability distributions of "payoffs" (sometimes referred to as "gambles" or "lotteries") into real numbers. Specifically, assuming the decision maker satisfies some axioms of rational behavior, a vNM utility function can be constructed for that person (firm) that has the property that the *expected utilities* (in a probabilistic sense) of gambles produce a *ranking* of the desirability of the gambles consistent with the decision maker's risk preferences. To illustrate their use, assume that a decision maker must face an alternative yielding two possible outcomes with payoffs, π_1 and π_2. Without loss of generality, assume $\pi_1 < \pi_2$. Suppose that outcome π_1 occurs with known probability p and π_2 with probability $(1 - p)$. The decision maker, in concert with the axioms of von Neumann-Morgenstern theory, evaluates his situation in terms of the expected utility. Let the situation be labeled, $\tilde{a}$ (the $\sim$ denotes randomness). Then $EU(\tilde{a}) = pU(\pi_1) + (1 - p)U(\pi_2)$. (Notice that the expected monetary value of the situation is $E(\tilde{a}) = p\pi_1 + (1 - p)\pi_2$.) If the decision maker is now presented with the choice between facing gamble (alternative) $\tilde{a}$, or some other gamble $\tilde{b}$ the decision maker can compare $EU(\tilde{a})$ with $EU(\tilde{b})$: the gamble producing the larger expected utility is the preferred gamble.

Many of the decision maker's risk-preference characteristics are captured by the mathematical properties of his utility function. The first characteristic is whether the decision maker (d.m.) would prefer facing situation $\tilde{a}$ with its uncertain outcome or receiving the amount $E(\tilde{a})$ with certainty; that is, is $EU(\tilde{a})$ $<$, $=$, or $>$ $U[E(\tilde{a})]$? It is generally assumed that $U[E(\tilde{a})] > EU(\tilde{a})$ for most decision makers. When this is the case, the decision maker is said to be *risk-averse*. A risk-averse utility function $U(\pi)$

has the properties the $U'(\pi) > 0$ and $U''(\pi) < 0$. This is illustrated in figure 5–1. In constructing the figure it was convenient to let $p \approx 0.5$. π^* is called the *certainty equivalent* of $\tilde{a}$. The decision maker whose preferences are represented by $U(\pi)$ would be indifferent between facing the uncertain situation, $\tilde{a}$, and receiving π^* with certainty, so $U(\pi^*) = EU(\tilde{a})$. The difference, $E(\tilde{a}) - \pi^*$, is called the "risk premium." The risk premium measures how much of the expected payoff a d.m. would be willing to forego to avoid the uncertainty associated with the gamble. For a risk-averse d.m. the risk premium is positive for all gambles, i.e., the d.m. would always be willing to accept less than the expected value of the gamble to eliminate the uncertainty. A *risk-seeking* d.m. has a utility function with the property that $EU(\tilde{a}) > U[E(\tilde{a})]$, so that its risk premium is always negative and the d.m. would be willing to "pay" more than $E(\tilde{a})$ in order to face the gamble $\tilde{a}$ itself. Finally, a *risk-neutral* d.m. has the property that $EU(\tilde{a}) = U[E(\tilde{a})]$. Such a d.m. essentially converts gambles into their expected monetary value for analysis—the risk premium is zero for all gambles. It is well-known that $U'' \lesseqgtr 0$ according as the d.m. is (risk-averse, risk-neutral, risk-seeking).

Most decision makers are considered to be risk-averse over a normal range of possible outcomes, but some may be more risk-averse than others. Since all utility functions that are positive linear transformations of one another (e.g., $\hat{U}(\pi) = a + bU(\pi)$ with $b > 0$) are strategically equivalent (i.e., they yield the same preference ranking among alternatives), the magnitude of $U''(\pi)$ alone is not a good measure of the degree of risk

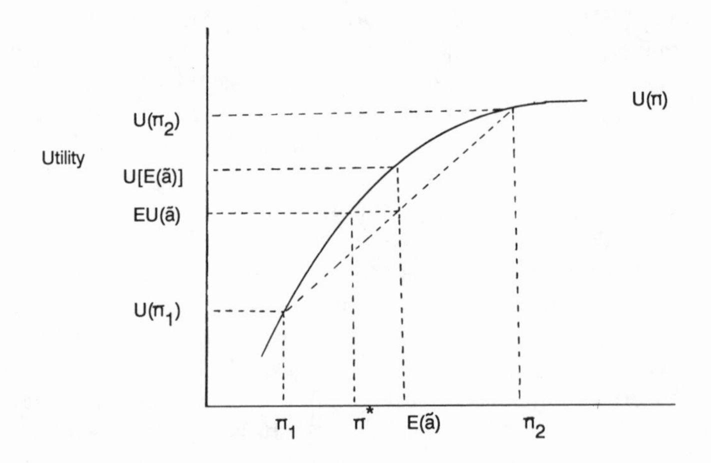

Figure 5–1. A Risk-Averse Utility Function for a Gamble $\tilde{a}$

aversion. Two strategically equivalent utility functions may have different values of $U''(\pi)$. Pratt [1964] has developed a measure of *absolute risk aversion*: $r(\pi) \equiv -U''(\pi)/U'(\pi)$. He has shown that two utility functions are strategically equivalent if and only if they have the same absolute risk aversion function. Furthermore, if $r(\pi)$ is positive for all π, then $U(\pi)$ is strictly concave ($U''(\pi) < 0$) and the decision maker is risk-averse. Finally, if the $r(\pi)$ function is known, the underlying utility function, $U(\pi)$ is known up to a positive linear transformation.

A decision maker is said to exhibit decreasing absolute risk aversion if $r(\pi)$ decreases as π increases ($r'(\pi) < 0$). This is consistent with behavior on the part of the decision maker whereby as the expected level of profit (or base wealth), $\bar{\pi}$, increases while the spread (variation) in payoffs around the mean is held constant, the d.m.'s risk premium ($E(\tilde{a}) - \pi^*$) for the gamble decreases. For example, consider a decision maker who faces a situation that could result in profit of $\pi_1 = 90$ or $\pi_2 = 110$ with $p = 0.5$ (so $E(\tilde{a}) = 100$). Compare this with the same decision maker who faces a situation that could result in profit π_1 of 990 or π_2 of 1010 with $p = 0.5$—so $E(\tilde{b}) = 1000$. The "gamble" is the same in each case (i.e., 50–50 chance of gaining or losing $10.00 from the expected profit or "base" profit level). It is generally thought that for most decision makers the risk premium is smaller in the second case; thus, $r'(\pi) < 0$. Examples of utility functions with decreasing absolute risk aversion are $U(\pi) = \ln(\pi + b)$ for $\pi \geqslant -b$, and $U(\pi) = \pi^{1-c}$ for $0 < c < 1$.

Another measure of risk aversion, called *proportional (or relative) risk aversion*, is also useful. The proportional risk-aversion function is $R(\pi) \equiv -\pi U''(\pi)/U'(\pi) = \pi r(\pi)$. The proportional risk aversion function measures the effect on the d.m.'s risk premium when the level of mean profit (or initial wealth) *and* the spread of the gamble increase in the same proportion. To illustrate, suppose the first situation faced by the decision maker has possible outcomes of $\pi_1 = 90$ or $\pi_2 = 110$ with $p = 0.5$ (mean or base profit of 100). The second situation faced by the same decision maker is $\pi_1 = 900$, $\pi_2 = 1100$ (mean profit of 1000). In the second situation the firm's mean profit and variation in profit around the mean have both increased by a factor of 10. If the risk premium for the d.m. in the second case is (greater than, equal to, or less than) 10 times the risk premium in the first case, the d.m. is said to exhibit (increasing, constant, or decreasing) proportional risk aversion; equivalently, $R'(\pi) \gtreqless 0$. It is usually assumed that $R'(\pi) \geqslant 0$. Examples of utility functions that exhibit increasing proportional risk aversion are $U(\pi) = \ln(\pi + b)$ and $U(\pi) = -\exp(-a\pi)$, while $U(\pi) = \pi^{1-c}$ exhibits constant proportional risk aversion. The functions $r(\pi)$ and $R(\pi)$ provide convenient summaries of important properties of $U(\pi)$ and we will often refer to $r(\pi)$, $R(\pi)$ and their derivatives in the following discussion of production/location problems on a line.

In the literature both assets and profits have been used as the arguments for vNM utility functions. Because the latter is more common and less cumbersome, throughout this book we will assume that the firm's utility function has been constructed in terms of profits, *given the firm's beginning wealth*. Any properties regarding $r(\pi)$ and $R(\pi)$ will be assumed to be with respect to (after-tax) profit.

Uncertainty in Nonspatial Production Analysis

Uncertainty has been introduced into nonspatial models of the firm in three ways: through uncertainty in the prices of—or demand for—output, uncertainty in input prices, and uncertainty in the production or technological relationships. In addition, there have been different assumptions invoked regarding what decisions must be made before the random phenomena are realized and how much ex post adjustment of the decision can be made by the firm. The following overview is in no way intended to be a complete review of the vast amount of published work on this topic; however, special attention should be directed to those models that incorporate uncertain prices, because it most closely parallels the production/location models in this chapter.

Uncertainty in Output Prices and Demand

The initial fundamental work incorporating output price uncertainty into the firm's production decision was performed by Baron [1970] and Sandmo [1971]. In their model the decision maker's utility function is represented by a vNM utility function of profit with $U'(\pi) > 0$ and $U''(\pi) < 0$. A cost function (rather than a production function illustrating inherent input substitutability) is assumed. If z_0 is the rate of output, and B is fixed cost, then total cost is

$$G(z_0) = C(z_0) + B \tag{5.1}$$

with $C(0) = 0$ and $C'(z_0) > 0$. If the price of the output, $\tilde{p}_0$, is a random variable (throughout this chapter the $\sim$ above a parameter or variable indicates that it is random) the profit function becomes

$$\tilde{\pi}(z_0) = \tilde{p}_0 z_0 - C(z_0) - B \tag{5.2}$$

with $E(\tilde{p}_0) = \bar{p}_0$ and realizations of $\tilde{p}_0$ all positive.

It is assumed that the first and second-order conditions determine a nonzero, finite, unique optimum for the problem of maximizing (5.2) with respect to z_0. Under price uncertainty, output is shown to be smaller than the output that would occur if the random parameter, $\tilde{p}_0$, were replaced by its expected value, $\bar{p}_0$, and the resulting certainty problem solved. More generally, Baron [1970] showed that the optimal output level is a nonincreasing function of the firm's absolute risk aversion, $r(\pi)$.

In addition to determining the direct effects of risk-averse preferences, Sandmo [1971] investigated the effects of marginal changes in the level of price uncertainty around the mean price. He performed his analysis by replacing $\tilde{p}_0$ with $\gamma\tilde{p}_0 + \theta$ (with $d\theta/d\gamma = -\bar{p}_0$). He could then perturb the mean price by just changing the value of θ, but the level of uncertainty was altered by increasing γ and simultaneously decreasing θ to maintain the same expected price. (This method of representing a random price and analyzing changes in the level of uncertainty seems unnecessarily cumbersome. Nevertheless, many subsequent authors have adopted this methodology for performing sensitivity analysis regarding changes in uncertainty. In this book we have adopted a simpler approach by defining $\tilde{p}_0 = \bar{p}_0 + \gamma\tilde{\varrho}$, where $\bar{p}_0 = E(\tilde{p}_0)$, $E(\tilde{\varrho}) = 0$, and $\gamma > 0$. Then changes in the level of uncertainty can be created by changing γ only, and changes in the expected price are represented by changes in $\bar{p}_0$.) Sandmo found that the sign of $dz_0^*/d\gamma$ is indeterminate for a general risk-averse utility function (* is used to denote optimal values). Batra and Ullah [1974], by invoking assumptions on the firm's production function, subsequently were able to show that an increase in uncertainty leads to a decline in optimal output (and factor demands), if the firm's absolute risk-aversion function is decreasing. Coes [1977] later established the same result without the restrictions on the production function.

Sandmo also investigated the impact of fixed costs. In the deterministic case, changes in the firm's costs, B, do not affect the optimal output rate. In contrast, under uncertainty with $r'(\pi) < 0$, an increase in fixed costs reduces the overall profit level, putting the decision maker at a point on his utility function where he is more risk-averse than would be the case with a smaller level of B. Consequently, z_0^* will be lower with larger levels of B.

Leland [1972] considered additional cases including where demand is uncertain and the firm can be both a price and quantity setter (i.e., a monopolist). The models of Baron and Sandmo, in which the firm is only a quantity setter, can be expressed as special cases of Leland's model. One of

Leland's main results is that the assumptions regarding which decisions must be made ex ante (before the uncertainty is resolved) affects the optimal behavior of the firm.

In the works discussed thus far it has been assumed that all decisions must be made before the realization of the random variables (i.e., ex ante). Turnovsky [1973] and Hartman [1976] modified these models by assuming a technology that allows ex post adjustments; that is, the firm selects the values of some variables (e.g., amount of capital) before the randomness is resolved, and then chooses the values of other variables (e.g., labor and output level) after seeing the realized output price. Epstein [1978] similarly allows some ex post adjustment and includes input price uncertainty of the "adjustable" inputs as well. Their main conclusions are that some of the earlier results of Sandmo and Batra and Ullah must be modified; mainly the effects of parametric perturbations become ambiguous because the production flexibility allowing ex post adjustment reduces the effects of the uncertainty. Pindyck [1982] includes uncertainty in costs as well as uncertainty in future demands in a multiperiod model.

The issue of ex post adjustment is especially relevant for, and recently has been introduced into, the production/location literature. This will be discussed later in this chapter; but the general study of dynamic attributes, including ex post adjustments, are an important area for future research in the context of production/investment/location problems and they are discussed further in the final chapter.

Uncertainty in Input Prices

Blair [1974] provided the first study of the impact of uncertainty in input prices. Letting the inverse demand function $p_0 = p_0(z_0)$ and the production function $z_0 = F(z_1, \ldots, z_n)$, the firm must commit itself to use some input rates, z_i, and produce some specified rate of output, z_0, prior to knowing the input prices which are random variables: $\tilde{p}_i = p_i(z_i, \tilde{u}) = \overline{p}_i(z_i) + \tilde{u}_i$ where $\overline{p}_i(z_i)$ is the "expected input supply function" and $\tilde{u}_i$ is a random variable with $E(\tilde{u}_i) = 0$.[1] The firm's problem is then to

$$\text{maximize} \quad E\left\{U\left[p_0(z_0)z_0 - \sum_{i=1}^{n} \tilde{p}_i(z_i, \tilde{u}_i)z_i\right]\right\} \tag{5.3}$$

where $z_0 = F(z_1, \ldots, z_n)$.

The first-order conditions for an optimum are

$$\partial E[U(\tilde{\pi})]/\partial z_i = E\{U'(\tilde{\pi})[MRP_i - (p_i + z_i \partial p_i/\partial z_i)]\} = 0$$
$$i = 1, \ldots, n \tag{5.4}$$

where $MRP_i = (p_0(z_0) + z_0 dp_0/dz_0)\partial z_0/\partial z_i$ is the marginal revenue product of input i. For risk-neutral or risk-averse firms the first- and second-order conditions are satisfied under the usual assumptions of diminishing marginal productivity. The second-order condition may be satisfied for the risk-averse firm even if marginal productivity is not declining.

Conditions (5.4) can be rewritten as

$$MRP_i - E[MFC_i] = \text{cov}[U'(\pi), MFC_i]/E[U'(\pi)]$$

or

$$MRP_i = E[MFC_i] + \text{cov}[U'(\pi), MFC_i]/E[U'(\pi)] \tag{5.5}$$

where $MFC_i = p_i + z_i \partial p_i/\partial z_i$ is the marginal factor cost of input i. Then, the effective marginal factor cost, which at an optimum is equal to MRP_i from (5.4), is equal to the *expected* marginal factor cost plus $\text{cov}[U'(\pi), MFC_i]/E[U'(\pi)]$. For a risk-averse firm, an increase in MFC_i would decrease π and increase $U'(\pi)$. So $\text{cov}[U'(\pi), MFC_i] > 0$ and the risk-averse firm therefore behaves as if marginal factor cost is higher than the expected marginal factor cost by an amount proportional to the covariance of $U'(\pi)$ and $(p_i + z_i \partial p_i/\partial z_i)$. This results in a lower output level and lower use of risky inputs than under certainty. Blair interprets this difference as the premium the firm is willing to pay to protect itself from uncertainty; thus, the risk-averse firm guards against excessive costs by employing fewer resources than the risk-neutral firm.

Blair also attempted to study the effects of shifts in the mean prices of inputs. However, even resorting to the two-input case, he concluded that the effect on the input levels is ambiguous. Using a two-input, fixed-output model where one input price is uncertain and the other is not, Martinich [1980] was able to eliminate Blair's ambiguous results by showing that an increase in the mean price of the risky input reduces its optimal use by a risk-averse firm. Using an identical two-input model, Stewart [1978] expanded on Blair's work to produce the following comparative-statics results: (1) Use of the risky input decreases (and of the riskless input increases) as the firm's risk aversion increases. (2) Use of the risky input increases as the variation in price around the mean (i.e., price uncertainty) decreases. Stewart further showed that these results are not affected by the inclusion of output price uncertainty, but if demand uncertainty is assumed, the results become ambiguous because demand and input price uncertainty create opposite effects.

The models and results in this subsection are used extensively in the stochastic production/location models presented later in this chapter.

Uncertainty in Technology

Of the many ways in which uncertainty can be introduced into the production relationship, perhaps the most interesting is to let the rate of production depend on flows of factor services randomly generated by the quantities of factors supplied. In a world of uncertainty, the flow of services from a given number of units of a factor may be a random variable. The randomness may be related to material quality, uncertain supply, the weather, quality control on output, machine breakdown, or deterioration of machine performance.

Technological uncertainty can be represented in various ways. For example, Feldstein [1971] assumed a Cobb-Douglas type of production function where

$$\tilde{z}_0 = \tilde{A} Z_1^{\tilde{b}} Z_2^{(1-\tilde{b})}$$

Uncertainty in this format can be interpreted as:

$$Z_1^{\tilde{b}} = Z_1^{\bar{b}} Z_2^{\gamma \tilde{\varrho}}$$

where $\bar{b} = E(\tilde{b})$ and $\tilde{b} = \bar{b} + \gamma \tilde{\varrho}$ with γ a constant. (Feldstein further greatly simplified his analysis by assuming that decision makers are risk-neutral.) Batra [1974] represented random output in a simpler, yet more general, form as

$$\tilde{z}_0 = \tilde{\alpha} F(Z_1, Z_2)$$

where a number of interpretations can be given to the random parameter, $\tilde{\alpha}$, including, for example, weather, strikes, and quality control.

An alternative approach (see Horowitz [1970]) differentiates the actual inputs from their productive flows of services. Specifically, consider a firm that produces a single output at rate z_0 using two inputs which are supplied in quantities Z_1 and Z_2 and which provide services at the rates z_1 and z_2. Then $z_1 = \tilde{u}_1 Z_1$ and $z_2 = \tilde{u}_2 Z_2$ where the $\tilde{u}_i$ are independently distributed random variables with unit means. The production function

$$\tilde{z}_0 = F(\tilde{z}_1, \tilde{z}_2) = F(\tilde{u}_1 Z_1, \tilde{u}_2 Z_2) \tag{5.6}$$

is assumed to be quasiconcave with $F_i > 0$, $i = 1, 2$ and $F(0, 0) = 0$. Here the random variables $(\tilde{u}_1, \tilde{u}_2)$ enter through the relatively complex (in contrast to price uncertainty) technological relationship, F. Costs are determined by the quantities Z_1 and Z_2 but production (and hence revenues) by $\tilde{z}_1$ and $\tilde{z}_2$. Ratti and Ullah [1976] use such a representation to analyze the impact of uncertainty in production on the competitive firm's demand for labor and capital (Z_1 and Z_2). To begin, Ratti and Ullah use Jensen's inequality to show that, if F is concave, then $E[F(\tilde{u}_1 Z_1, \tilde{u}_2 Z_2)] < F(Z_1, Z_2)$.[2] That is, the expected output generated by a given input factor combination (Z_1, Z_2) is less than the output that would be given by the same input combination when the random variables $\tilde{u}_1$ and $\tilde{u}_2$ are replaced by their expected values $\bar{u}_1 = 1$, $\bar{u}_2 = 1$. Notice that this conclusion does not depend on the properties of the firm's utility function.

Assuming that the firm seeks to maximize the expected utility of profits, Ratti and Ullah then formulate the firm's problem as

$$\text{maximize} \qquad EU(\tilde{\pi}) = EU[p_0 F(\tilde{u}_1 Z_1, \tilde{u}_2 Z_2) - p_1 Z_1 - p_2 Z_2] \quad (5.7)$$

for which the usual first and second-order conditions can be derived. The firm's usage of inputs under uncertainty can be contrasted with its usage under certainty. Using what the authors consider to be plausible assumptions (including $F_i > 0$, $F_{ii} < 0$, $F_{ij} > 0$, $i, j = 1, 2$) they show: (a) the risk-averse (prone) firm demands less (more) of both inputs than the risk-neutral firm and (b) the risk-neutral firm demands less of both inputs than it would under certainty.

In order to demonstrate the first part of their result, the authors are forced to invoke an unusual assumption concerning the curvature properties of the production function—that the elasticity of the marginal-product curves, N_i, is greater than minus one—i.e., that

$$N_i = [z_i(\partial^2 F/\partial z_i^2)][\partial F/\partial z_i]^{-1} > -1.$$

(Notice the similarity between N_i and $R(\pi)$.) With this assumption on N_i, it is shown that the expected value of the marginal revenue product of an input is greater than the input price at an optimum for risk-averse decision makers. Thus, the firm demands less (recall the assumption of diminishing marginal productivity) of the input than does a risk-neutral firm for whom expected marginal revenue product equals input price at an optimum. To achieve the second result, which compares input use of a risk-neutral firm with a certainty ($\tilde{u}_1 = \tilde{u}_2 = 1$) situation, further assumptions on the production function, beyond $N_i > -1$, are required. Note that when uncertainty is introduced through prices, the optimal choices of a risk-neutral firm under certainty and uncertainty are the same. Further, note that production uncertainty becomes equivalent to product price uncertainty whenever there is multiplicative technological uncertainty of the form

$$F(\tilde{u}_1 Z_1, \tilde{u}_2 Z_2) = L(\tilde{u}_1, \tilde{u}_2) F(Z_1, Z_2)$$

One of the key assumptions in the foregoing presentation is the so-called proportionality assumption: $\tilde{z}_i = \tilde{u}_i Z_i$. Pope and Just [1978] have argued that this is unrealistic, particularly with respect to the capital input, say Z_2. When the capital stock is increased by a constant, c, the standard deviation of the services, $\tilde{z}_2$, is also increased by a factor, c. This is not consistent with the notion that extra capital, in the form of back-up equipment, reduces the variability of service flows.

Ratti [1978] has responded that traditional economics incorporates the assumption that all inputs are fully utilized. The back-up equipment concept can be incorporated if the rate of utilization of capital, k, is included in the model as a variable. Then $\tilde{z}_2 = \tilde{u}_2 k Z_2$. Using this extended model, Ratti concludes that the risk-averse firm demands less of each input and chooses a lower rate of capital input utilization than the risk-neutral firm.

Pope and Kramer [1979] stress the importance of the marginal effects of input use on the probability distribution of output. In their presentation, the production function is: $F(z_1, \ldots, z_n, \tilde{e})$, where $\tilde{e}$ is a random disturbance. An input, j, is defined to be marginally risk increasing [decreasing] as covariance $[U'(\pi), F_j] < [>] \, 0$. With decreasing marginal productivity, an increase in z_j (ceteris paribus) leads to a decrease in F_j. If the covariance term is <0, then the slope of the utility function, $U'(\pi)$, must be increasing and so profit, π, must be decreasing. Then an input is marginally risk-reducing (-increasing) if the risk-averse firm uses a larger (smaller) quantity of the input than a risk-neutral firm. When the production function can be written as $F(z_1, \ldots, z_n, \tilde{e}) = g(\tilde{e}) f(z_1, \ldots, z_n)$, all inputs are marginally risk-increasing.

Alternative objectives have been used in models of the firm under uncertainty. We present just one example to illustrate the wide range of thought that exists in this area. Stewart [1982] has proposed that firms (owners) and their managers employ different definitions of risk. Managers perceive that outcomes leading to higher profits generate but modest additional rewards while outcomes resulting in exceptionally low profits may lead to unemployment. Thus, following Fishburn [1977], decision makers often associate risk with failure to obtain some target level of profit, $\hat{\pi}$. Then the key factor for managers is the probability distribution of $\hat{\pi} - \tilde{\pi}$ or the probability that the actual profit will be $\hat{\pi} - \pi$ units below the target level. For values of profit above $\hat{\pi}$ the manager's utility function is $U(\pi) = \pi$; for values of profit below $\hat{\pi}$ the manager's utility function is $U(\pi) = k\varphi(\hat{\pi} - \pi)$, where k weights the probability, φ, that the target profit is missed by $\hat{\pi} - \pi$ units. Not surprisingly, this approach leads to different results than those presented above. If this type of utility function represents actual preferences better than those used above, models using such an approach are worthy of future attention.

The material contained in this section is a representative sampling of the nonspatial economic theory of the firm under three kinds of uncertainty. The discussion in the following sections treats uncertainty in input prices and product prices, in transport rates, and in technology in a spatial context with a line as the location space. As illustrated above, uncertainty in production is an interesting case. Unfortunately, very little work has been done on technological uncertainty in a spatial context.

Uncertainty in Production/Location Problems

Following the structure of chapter 2, the analysis will again be divided into two groups of problems: problems in which the output transport costs are ignored and those that explicitly include output transport costs. Although uncertainty can be introduced in many ways, as illustrated in the preceding section, most of the discussion and analysis will focus upon the case where randomness occurs in the prices. This is because most published research has been restricted to this case [Martinich and Hurter 1982, 1985; Mathur 1983, 1985; Hsu and Mai 1984; Hurter and Martinich 1985; Mai 1984a, 1984b, Shieh [1987]]. The extent to which these results can be generalized to randomness in transport rates and technology is developed later.

Model Formulations and Example with Price Uncertainty

The models in this chapter are a direct generalization of those in chapter 2, so the same notation, location space, and assumptions will be used unless otherwise stated. In summary then, the firm produces one output at the rate z_0 using two inputs at rates z_1 and z_2 according to a quasiconcave production function, F. The sources of inputs 1 and 2 are at M_1 and M_2, as illustrated in figure 2–1; the distance between them is s. The location of the firm's plant, x, is measured by its distance from M_1; and $c_i(x)$ is the per unit cost of transporting input i from M_i to x. Initially it is assumed that the cost of transporting output is nill, so initially, the location of the market is unimportant. The primary change introduced here is that the prices for the inputs at their source locations and the price of the output are random variables, designated $\tilde{p}_1$, $\tilde{p}_2$, and $\tilde{p}_0$, respectively. The firm is assumed to have preferences toward risk, which are expressed by a vNM utility function, U. In the following models it is assumed that the firm must choose its plant location and input levels (and thus its output level) before the random variables are realized; this is to be done so as to maximize the firm's expected utility of profit.

The firm's stochastic production/location problem, corresponding to (PL2.1), can then be written as

(PL5.1)

$$\text{maximize} \quad EU[\tilde{p}_0 F(z_1, z_2) - (\tilde{p}_1 + c_1(x))z_1 - (\tilde{p}_2 + c_2(x))z_2] \tag{5.8}$$

$$\text{subject to} \quad z_1 \geqslant 0,\ z_2 \geqslant 0 \tag{5.9}$$

$$0 \leqslant x \leqslant s \tag{5.10}$$

If the output level is set a priori at $\bar{z}_0$ the firm's stochastic design/location problem, corresponding to (DL2.1), becomes

(DL5.1)

$$\text{maximize} \quad EU[\tilde{p}_0 \bar{z}_0 - (\tilde{p}_1 + c_1(x))z_1 - (\tilde{p}_2 + c_2(x))z_2] \tag{5.11}$$

$$\text{subject to} \quad \bar{z}_0 = F(z_1, z_2) \tag{5.12}$$

$$z_1 \geqslant 0,\ z_2 \geqslant 0 \tag{5.9}$$

$$0 \leqslant x \leqslant s \tag{5.10}$$

(Notice that the argument of U in (5.11) is random profit rather than cost, as used in the deterministic models; this is because it is most convenient to have utility expressed as a function of profit for all models.)

Output transport costs can be incorporated into these models in a fashion similar to chapter 2. The output market is assumed to be located at M_2 and the per unit transport cost from the plant location, x, to M_2 is $c_0(x)$. The firm's stochastic production/location problem is then

(PL5.2)

$$\text{maximize} \quad EU[(\tilde{p}_0 - c_0(x))F(z_1, z_2) - (\tilde{p}_1 + c_1(x))z_1 - (\tilde{p}_2 + c_2(x))z_2] \tag{5.13}$$

$$\text{subject to} \quad z_1 \geqslant 0,\ z_2 \geqslant 0 \tag{5.9}$$

$$0 \leqslant x \leqslant s \tag{5.10}$$

If output level is fixed a priori at $\bar{z}_0$, the firm's stochastic design/location problem becomes

$$\text{maximize} \quad EU[(\tilde{p}_0 - c_0(x))\bar{z}_0 - (\tilde{p}_1 + c_1(x))z_1 - (\tilde{p}_2 + c_2(x))z_2] \qquad (5.14)$$

$$\text{(DL5.2)} \quad \text{subject to} \quad z_1 \geqslant 0,\ z_2 \geqslant 0 \qquad (5.9)$$

$$0 \leqslant x \leqslant s \qquad (5.10)$$

$$\bar{z}_0 = F(z_1, z_2) \qquad (5.12)$$

The following example illustrates the possible effects of introducing price uncertainty and general risk preferences into the production/location framework.

Example 5.1

Consider problem (DL5.1) with $F(z_1, z_2) = (z_1 z_2)^{0.5}$; $\bar{z}_0 = 4$; $p_0 = 8$, $p_2 = 4$, and $\tilde{p}_1 = 3.5 - \tilde{\varrho}$ where $\text{Prob}(\tilde{\varrho} = n) = 0.2$ for $n = -2, -1, 0, 1, 2$ (that is, only the price of the first input is random—the price of the output and second input are deterministic). Also let $s = 1$, $c_1(x) = 0.5x$, $c_2(x) = 0.5 - 0.5x$. The firm's utility function is assumed to be $U(\pi) = -e^{-a\pi} = -\exp(-0.1\pi)$; so $r(\pi) = -U''/U' = 0.1$, $r'(\pi) = 0$, $R(\pi) = 0.1\pi$, $R'(\pi) = 0.1$.

The optimal solution for this problem is $x^* = 1$, $z_1^* = 3.691$, $z_2^* = 4.335$. The optimal solution for the same problem, but with $\tilde{p}_1$ replaced with its expected value of 3.5 is $x^* = 0$, $z_1^* = 4.536$, $z_2^* = 3.527$. Thus, the inclusion of uncertainty and risk aversion alters the firm's optimal location and production decision. Specifically, the firm responds to the uncertainty (vis a vis the deterministic case) by reducing its use of the risky input and increasing its use of the price-certain input; this causes the pull of the price-certain input to overcome that of the price-uncertain input and shifts the optimal location to the source of the price-certain input. The effects of uncertainty and risk preferences on the firm's optimal solution will be explored in depth later in the chapter. A second issue, however, is brought out by this example: endpoint optimality. We will now identify those conditions under which (only) endpoints can be optimal.

Endpoint Optimality under Price Uncertainty

Although the inclusion of price uncertainty and general risk preferences generalizes and complicates the deterministic models of chapter 2, the endpoint optimality properties for the stochastic-price case are identical to those for the deterministic case. This will be shown in some detail below. Throughout the following discussion it is assumed that a finite optimum exists for the problem under consideration.

Theorem 5.1

If all transport costs for the inputs and output are concave, but not linear, in distance (with "jumps" allowed at the source of the shipment), then the optimal plant locations for problems (PL5.1), (DL5.1), (PL5.2), and (DL5.2) can *only* be at endpoints of $[0, s]$.

Proof: Since any optimal solution represents a "shipping pattern" the result follows immediately from Wendell and Hurter [1973b]. □

In a similar way, when transport costs are linear in distance shipped, Hakimi's [1964] result that an endpoint location will be optimal easily extends to this case. This result can be strengthened for (DL5.1) just as it was in chapter 2 for (DL2.1). We begin by constructing the Kuhn-Tucker conditions necessary for an optimum; these will be used throughout the subsequent discussion.

The Lagrangian function for (DL5.1) is

$$L = EU\{p_0\bar{z}_0 - [\tilde{p}_1 + c_1(x)]z_1 - [\tilde{p}_2 + c_2(x)]z_2\} + \lambda[\bar{z}_0 - F(z_1, z_2)] + \mu(s - x) \tag{5.15}$$

The Kuhn-Tucker conditions are then

$$\partial L/\partial z_i = E[U'(\tilde{\pi})[-\tilde{p}_i - c_i(x)] - \lambda^* F_i \leqslant 0 \qquad i = 1, 2 \tag{5.16}$$

$$z_i^*(\partial L/\partial z_i) = 0 \qquad i = 1, 2 \tag{5.16a}$$

$$\partial L/\partial x = E[U'(\tilde{\pi})[-c_1'(x)z_1^* - c_2'(x)z_2^*] - \mu^* \leqslant 0 \tag{5.17}$$

$$x^*(\partial L/\partial x) = 0 \tag{5.17a}$$

$$\partial L/\partial \lambda = \bar{z}_0 - F(z_1^*, z_2^*) = 0 \tag{5.18}$$

$$\partial L/\partial \mu = s - x^* \geqslant 0 \tag{5.19}$$

$$\mu^*(\partial L/\partial \mu) = 0 \tag{5.19a}$$

$$\lambda^* \leqslant 0, \ \mu^* \geqslant 0 \tag{5.20}$$

The asterisk (*) denotes optimal values.

Condition (5.17) is equivalent to condition (2.11) for (DL2.1); in fact, it identifies the same three possible cases: (i) If $c_1'(x^*)z_1^* + c_2'(x^*)z_2^* > 0$, then $x^* = 0$; (ii) If $c_1'(x^*)z_1^* + c_2'(x^*)z_2^* < 0$, then $x^* = s$; thus, (iii) From (i) and (ii), $x \in (0, s)$ only if $c_1'(x^*)z_1^* + c_2'(x^*)z_2^* = 0$.

The expression $c_1'(x^*)z_1^* + c_2'(x^*)z_2^*$, which is free of random variables in this model, is the net locational pull toward zero (M_1). (Recall, $c_1'(x) > 0$ while $c_2'(x) < 0$.) Just as for the deterministic model, when it is positive, the pull of the first input exceeds that of the second and $x^* = 0$; when the NLP is negative the opposite occurs and $x^* = s$. Only when the pulls from the two input locations exactly balance could an interior point of $[0, s]$ be optimal. Showing that this cannot occur is at the heart of the following.

Theorem 5.2

If transport costs for inputs are concave in distance (with "jumps" allowed at $x = 0$ and $x = s$), then the optimal plant location for (DL5.1) can only be at an endpoint of the line segment $[0, s]$.

Proof: It follows from Wendell and Hurter [1973b] that an interior location can be optimal only if the $c_i(x)$ are both linear in x over $[0, s]$. Consequently, our proof must deal only with $c_1(x) = r_1 x$ and $c_2(x) = r_2(s - x)$. The remainder of the proof, which is by contradiction, follows the argument used for theorem 2.2.

Consider $c_1'(x^*)z_1^* + c_2'(x^*)z_2^* = r_1 z_1^* - r_2 z_2^* = 0$. Then the optimal level of expected utility is:

$$EU[p_0\bar{z}_0 - \tilde{p}_1 z_1^* - r_1 x z_1^* - \tilde{p}_2 z_2^* - r_2(s - x)z_2^*]$$

$$= EU[p_0\bar{z}_0 - \tilde{p}_1 z_1^* - \tilde{p}_2 z_2^* - r_2 s z_2^*]$$

which is independent of x. Therefore, for *any* and all locations $\bar{x} \in [0, s]$, (z_1^*, z_2^*) must satisfy the optimality conditions (5.16) rewritten as

$$\frac{E[U'(\tilde{\pi})(-\tilde{p}_1 - r_1\bar{x})]}{E[U'(\tilde{\pi})(-\tilde{p}_2 - r_2(s - \bar{x}))]} = \frac{F_1(z_1^*, z_2^*)}{F_2(z_1^*, z_2^*)} \tag{5.21}$$

The right-hand side of (5.21) is independent of $\bar{x}$ while the numerator of the left-hand side is decreasing and the denominator is increasing in $\bar{x}$. Consequently, (5.21) can hold at most for a single location $\bar{x} \in [0, s]$ and not for all locations—a contradiction. □

This result can be extended from the design/location context to the production/location context where the rate of output, z_0, is also a variable.

Theorem 5.3

If transport costs are concave in distance (with "jumps" allowed at $x = 0$, $x = s$), then the optimal plant location for (PL5.1) can occur only at an endpoint of $[0, s]$.

Proof: Let z_0^* be the optimal level of output for (PL5.1). For this level of output, the choice of input rates and location must maximize expected utility; that is, they must optimize (DL5.1). The result then follows immediately. □

Theorems 5.1–5.3 generalize and extend the results for the deterministic cases presented in chapter 2. These theorems, which restrict the search for optimal solutions to endpoints, hold for risk-averse or risk-prone utility functions as well as for production relationships with increasing or decreasing returns to scale. The two crucial features of these models are the substitutability of the inputs in the production process and the fact that the random component does not directly influence (multiply) the location variable. Although the inclusion of price uncertainty and general risk preferences can change the site of the optimal location, under the given conditions, the optimal plant location must still be at an endpoint.

The extension of deterministic endpoint-optimality results to the stochastic case is completed by the following.

Theorem 5.4

Suppose that the transport cost functions in (DL5.2) and (PL5.2) are concave in distance (with appropriate jumps allowed at the endpoints). If $z_2^* > 0$, then the optimal plant location can only be at an endpoint of $[0, s]$.

Proof: From theorem 5.1 and Wendell and Hurter [1973b] it is sufficient to prove this for the case where the transport costs are linear in distance. The proof of this is essentially identical to that used for theorems 2.4 and 2.5, and is omitted for brevity. □

As with the deterministic models, if one or more of the transport cost functions is not concave, then it is possible that a location on the interior of $[0, s]$ can be optimal. Examples of this would include transmission of some energy inputs (or output) that have superlinear transmission losses with distance transmitted.

Endpoint Optimality under Other Forms of Parametric Uncertainty

In this subsection we identify whether the previous endpoint-optimality results generalize when uncertainty is incorporated in the model in other ways. To simplify the discussion it will be assumed that transport costs are all linear in distance shipped so that the expression for profit becomes

$$\pi = (p_0 - r_0(s - x))z_0 - (p_1 + r_1 x)z_1 - (p_2 + r_2(s - x))z_2 \quad (5.22)$$

where the source of the uncertainty is left vague for now. Given the general structure of our models, uncertainty in output demand is usually assumed to be captured through price uncertainty (although this is not always the case in a monopolistic environment). Consequently, here we will focus upon uncertainty in the production process and in transport rates.

Perhaps the most straightforward way to include uncertainty in the production process is to introduce a random variable, $\tilde{e}$, and write

$$\tilde{z}_0 = \tilde{e}F(z_1, z_2) \quad (5.23)$$

As indicated earlier in this chapter, there are many alternative ways of including technological uncertainty. This simple approach will suffice here since our purpose is to explore the robustness of the node optimality result under alternative forms of uncertainty.

Keeping the problem at a general level for now, we do not specify which parameters are uncertain, and do not fix the output level a priori. The firm's stochastic production/location problem is then to

$$\text{maximize} \qquad EU[\tilde{\pi}] \quad (5.24)$$

subject to (5.22), (5.23), (5.9), and (5.10). The primary determinant of generalized endpoint optimality is the way in which the probability distribution of $\tilde{\pi}$ depends upon the plant location, x. Now consider $E[U(\pi)]$ to be a function of the variables z_1, z_2, and x (with $eF(z_1, z_2)$ substituted for z_0). If we select an arbitrary value for the location variable, x, the resulting optimization problem can be solved for optimal values of the input variables z_1 and z_2. The first-order conditions for this optimization yield the optimal values of z_1 and z_2 parameterized on x; we denote these solutions as $z_1(x)$ and $z_2(x)$. Substituting these functions for z_1 and z_2 in (5.22), then substituting (5.22) and (5.23) into (5.24), and then optimizing with respect to x yields

$$dEU(\tilde{\pi})/dx = \sum_{i=1}^{2} [(\partial EU(\tilde{\pi})/\partial z_i(x))(dz_i(x)/dx)] + \partial EU(\tilde{\pi})/\partial x \quad (5.25)$$

By definition, $z_i(x)$, as the optimal value of z_i at an arbitrary fixed x, must satisfy the first-order conditions $\partial EU(\tilde{\pi})/\partial z_i^* = 0$. Therefore,

$$dEU(\tilde{\pi})/dx = \partial EU(\tilde{\pi})/\partial x \tag{5.26}$$

But

$$\partial EU(\tilde{\pi})/\partial x = E[U'(\tilde{\pi})(\partial\pi/\partial x)] \tag{5.27}$$

where

$$\partial\pi/\partial x = r_0 eF(z_1(x), z_2(x)) - r_1 z_1(x) + r_2 z_2(x) \tag{5.28}$$

For an interior optimal location to occur (5.27) must equal zero at $x = x^*$.

The expression (5.27) is the analogue of (5.17) in our earlier model. In that model, the uncertainty arose through the assumed randomness of the price parameters p_0, p_1, and p_2. These parameters do not appear in (5.28); thus, $\partial\pi/\partial x$ is deterministic, so that (5.27) can be rewritten as $EU'[\tilde{\pi}](\partial\pi/\partial x)$. Under those conditions, the interpretation of $\partial\pi/\partial x = 0$ is straightforward, and theorems 5.2–5.3 could be proven by showing $\partial\pi/\partial x = 0$ is not possible and so only endpoints can be optimal locations.

However, when the uncertainty arises from randomness in the transport rates, r_0, r_1, r_2, or the production process through e, the interpretation of $\partial\pi/\partial x = 0$ is not straightforward since the partial derivative, itself, is a random variable. This means that $\partial\pi/\partial x$ cannot be removed from inside the expectation operator in (5.27), and the proofs used in the preceding section do not apply. If the firm is not risk-neutral (i.e., $U'' \neq 0$) an optimum interior location is possible; that is, (5.27) can equal zero. This depends on whether or not the risk is "location dependent." That is, *for fixed values of the variables*, if the variance of $\tilde{\pi}^*$ with respect to x is constant, then maximizing $EU(\tilde{\pi})$ (with respect to the location, x) is accomplished at the location with the largest expected profit—namely, one of the nodes. However, when the variance of profit depends on the plant location, there is room for trade-off between expected profit and variance of profit in the maximization of $EU(\pi)$ for a risk averse decision maker and interior ($0 < x < s$) optimal locations are possible. This might occur if, for example, a node location is associated with a greater degree of risk (i.e., larger variance) than are interior locations of $[0, s]$.

Whether the variance of profit is or is not independent of the location, x, can be determined by the randomness of $\partial\pi/\partial x$ as given by (5.28). Following Park and Mathur [1986a] we will show that the variance of profit is independent of location only if $\partial\pi/\partial x$ is not random. Recall that in the formulation (5.28) the optimal values of z_1 and z_2 are employed for each value of x.

$$\begin{aligned} \operatorname{var} \tilde{\pi} &= E(\pi^2) - (E\pi)^2 \\ \partial(\operatorname{var} \pi)/\partial x &= E[2\pi(\partial\pi/\partial x)] - 2E(\pi)E(\partial\pi/\partial x) \\ &= 2\operatorname{cov}(\pi, \partial\pi/\partial x) \end{aligned}$$

Whenever $\partial\pi/\partial x$ is not random, then $\operatorname{cov}(\pi, \partial\pi/\partial x) = 0$ and the variance of profit is independent of the plant location. We have already indicated that $\partial\pi/\partial x$ is random whenever any or all of the parameters r_0, r_1, r_2, and e are random and that $\partial\pi/\partial x$ is *not* random when only some or all of p_0, p_1, and p_2 are random. Thus, node optimality holds when uncertainty enters through input and/or output prices and transport costs are concave in distance. It *may not* hold when uncertainty enters through transport costs and/or technological parameters. To illustrate this possibility, we present the following example where there is uncertainty in the transport rates.

Example 5.2

Consider problem (DL5.1) with $F(z_1, z_2) = (z_1 z_2)^{0.5}$; $\bar{z}_0 = 4$; $p_0 = 20$, $p_1 = p_2 = 4$; $s = 1$. The transport rates are assumed to be (highly correlated) random variables: $c_1(x) = \tilde{r}_1 x$ and $c_2(x) = \tilde{r}_2(1 - x)$ where $\operatorname{Prob}(r_1 = 1, r_2 = 6) = 0.5$ and $\operatorname{Prob}(r_1 = 6, r_2 = 1) = 0.5$. The firm's utility function is assumed to be $U(\pi) = -\exp(-0.1\pi)$.

The optimal solutions for this problem are $x^* = 0.22$, $z_1^* = 5.07$, $z_2^* = 3.16$, $EU = -0.03330$, and the complement of this solution, $x^* = 0.78$, $z_1^* = 3.16$, $z_2^* = 5.07$, $EU = -0.03330$. The optimal production mix at $x = 0$ is $z_1(0) = 5.98$ and $z_2(0) = 2.68$ with $EU = -0.03375$ (the solution at $x = 1$ has the values of z_1 and z_2 reversed). The expected profit at the optimal location is actually less than at the endpoints (34.58 vs. 35.98), but the variation in profit is much less because the negative correlation of the transport rates tends to reduce the risk.

The preceding results are consistent with the results of Alperovich and Katz [1983]. Using a model with a single input and linear production function, they concluded that optimal interior locations are possible when the transport rate for the single product is a random variable and $U'' < 0$.

It is important to note that a crucial condition for the above results is that the utility function be nonlinear, i.e., not risk-neutral. If the firm is risk-neutral, then U' is a constant and condition (5.27) becomes

$$E[U'(\pi)(\partial\pi/\partial x)] = cE(\partial\pi/\partial x) \tag{5.29}$$

where $c = U'(\pi)$. Then $E(\partial\pi/\partial x)$ must equal zero for an interior location to be optimal. But the same argument as that used for theorem 5.2 can be used with only minor changes to prove that this will never occur. Therefore, for a risk-neutral firm endpoint optimality is ensured *even when uncertainty enters through the transport rates and technology.*

It has been assumed that the firm must select the values of both the production and location variables before the realization of any random variables. A few models have appeared recently which only require the location variable to be selected before, while the output and input levels can be selected after, the realization of the random variables. Mai [1981] used this convention with uncertain output price and risk neutrality, but he did not investigate the issue of endpoint optimality. Instead, he assumed the optimal location was an interior point of the line and studied the effects of uncertainty on the optimal location. Mai [1984a] extended his work to the case of imperfect competition while retaining the risk neutrality and interior optimal location assumptions. Park and Mathur [1986b] have considered production uncertainty in this format. In evaluating a location, x, the optimal input levels at that location, $z_1(x)$ and $z_2(x)$, may be random variables themselves because the optimal input levels may change as a result of the realization of the random parameter. Park and Mathur show that if the production uncertainty is multiplicative and F is homothetic, then $z_1(x)/z_2(x)$ is nonrandom ex ante, so that the NLP is nonrandom ex ante and interior plant locations cannot be optimal. Where F is not homothetic, however, the optimal input ratio at each location becomes random and an interior optimal location becomes possible. Although it has not been proved formally, it appears that a similar result would hold for output price uncertainty. In addition, with ex post adjustments, input price uncertainty would appear to allow interior optimal locations even if F is homothetic because the optimal input ratio at each location would depend on the price realizations of the inputs. In each of these cases, however, interior optimal locations are impossible if the firm is risk-neutral, unless, of course, the transport costs are nonconcave. (A somewhat different two-state production/location model in a network space, by Louveaux and Thisse [1985], is discussed in chapter 6.)

There are almost certainly problem scenarios where the two-stage model is the more appropriate just as there are planning-problem scenarios where (at least tentative) decisions on all decision variables must be made prior to the realization of the random variables. We have adopted the latter modeling approach in this volume because of its more common use in the literature. For that reason we are unable to do full justice to Park and Mathur's stimulating efforts here; however, we feel that two-stage modeling should be an area of additional research.

Parametric Analysis: Design/Location Problems

Introduction and Example

The inclusion of price uncertainty and risk preferences will normally alter the optimal input mix and location in comparison with the deterministic version of design/location problems (DL5.1). The effects of differences in the level of uncertainty and the degree of risk aversion on the firm's optimal decision can best be understood by first studying a numerical example. For illustrative purposes, except where specified otherwise, in the remainder of this chapter uncertainty will be restricted to the price of the first input which will be written: $\tilde{p}_1 = \bar{p}_1 - \gamma\tilde{\varrho}$ where $E(\tilde{\varrho}) = 0$ and $\gamma > 0$. The parameter γ is used to control the extent of uncertainty or price variation in the first input. All other prices are assumed to be known with certainty so $\tilde{p}_0 = p_0$ and $\tilde{p}_2 = p_2$. Differences in firms' aversion to risk can be characterized by their absolute risk aversion functions $r(\pi)$. Of two firms, A and B, with von Neumann-Morgenstern utility functions, U_a and U_b, B is the more risk-averse if $r_b(\pi) \geqslant r_a(\pi)$ for all π with strict inequality holding for some π in every interval.

Example 5.3

Consider a generalization of example 5.1. Specifically, let $F(z_1, z_2) = (z_1 z_2)^{0.5}$; $\bar{z}_0 = 4$; $p_0 = 8$, $p_2 = 4$; $s = 1$, $c_1(x) = 0.5x$, $c_2(x) = 0.5 - 0.5x$. The only changes are that $\tilde{p}_1 = 3.5 - \gamma\tilde{\varrho}$ where $\text{Prob}(\tilde{\varrho} = n) = 0.2$ for $n = -2, -1, 0, 1, 2$ and $\gamma > 0$; and the utility function is written more generally as

$$U(\pi) = \begin{cases} -\exp(-a\pi) & \text{for } a > 0 \\ \pi & \text{when } a = 0 \end{cases}$$

Notice that $r(\pi) = -U''/U' = a$; $r'(\pi) = 0$; $R(\pi) = a\pi$; $R'(\pi) = a > 0$. Thus, firms with utility functions in this family exhibit *constant* absolute risk aversion and increasing proportional risk aversion. Increasing values of a indicate utility functions with greater absolute risk aversion. Also notice that increasing values of γ indicate greater price uncertainty.

The optimal solutions for this problem are given in table 5–1 for four utility functions ($a = 0, 0.05, 0.10, 0.20$) and for seven levels of price uncertainty ($\gamma = 0, 0.2, 0.4, 0.6, 0.8, 1.0, 1.2$). To illustrate the effects of increasing levels of risk aversion, consider the cases where $\gamma = 1$, so $\tilde{p}_1 = 3.5 - \tilde{\varrho}$. For the risk-neutral firm ($a = 0$), the optimal solution is $z_1^* = 4.536$, $z_2^* = 3.527$, $x^* = 0$. For the firm with $a = 0.05$: $z_1^* = 4.286$,

Table 5–1. Optimal Input Mixes and Location for Example 5.3[a]

a	γ	$x = 0$: z_1^*	$x = 0$: z_2^*	$x = 0$: EU	$x = s$: z_1^*	$x = s$: z_2^*	$x = s$: EU
0.05	0.2	4.524	3.537	−0.990[b]	3.992	4.008	−1.002
	0.4	4.490	3.564	−0.996[b]	3.969	4.031	−1.006
	0.6	4.436	3.607	−1.005[b]	3.931	4.070	−1.014
	0.8	4.366	3.665	−1.019[b]	3.882	4.122	−1.025
	1.0	4.286	3.733	−1.036[b]	3.824	4.184	−1.039
	1.2	4.196	3.813	−1.056	3.759	4.256	−1.055[b]
0.10	0.2	4.512	3.546	−0.983[b]	3.984	4.016	−1.006
	0.4	4.448	3.597	−1.007[b]	3.940	4.061	−1.025
	0.6	4.350	3.687	−1.046[b]	3.870	4.134	−1.056
	0.8	4.234	3.779	−1.099[b]	3.785	4.227	−1.099
	1.0	4.108	3.895	−1.166	3.691	4.335	−1.153[b]
	1.2	3.980	4.020	−1.246	3.592	4.454	−1.218[b]
0.20	0.2	4.490	3.564	−0.982[b]	3.969	4.031	−1.026
	0.4	4.372	3.660	−1.076[b]	3.886	4.117	−1.102
	0.6	4.218	3.793	−1.233	3.771	4.243	−1.231[b]
	0.8	4.052	3.949	−1.459	3.644	4.391	−1.413[b]
	1.0	3.892	4.111	−1.761	3.517	4.549	−1.654[b]
	1.2	3.742	4.276	−2.152	3.396	4.711	−1.962[b]
$a = 0$ or $\gamma = 0$		4.536	3.527	—[b]	4.000	4.000	

[a] The optimal input mix and corresponding expected utility is listed for $x = 0$ and $x = s = 1$.

[b] Global optimum for each value of a and γ.

$z_2^* = 3.733$, $x^* = 0$. For the firm with $a = 0.10$: $z_1^* = 3.691$, $z_2^* = 4.335$, $x^* = s$. Finally, for the firm with $a = 0.20$: $z_1^* = 3.517$, $z_2^* = 4.549$, $x^* = s$.

A pattern is suggested by these results: more risk-averse firms use less of the uncertain-price input, input 1. Because the output level is assumed fixed, as z_1 decreases, z_2 must increase. This results in a tendency for more risk-averse firms to locate at the source of the second input, whereas less risk-averse firms use more of the uncertain-price input and locate at its source. The underlying rationale for such behavior is that risk-averse firms behave as if uncertainty in an input price adds an additional "risk premium" to the expected price of the input. The more risk-averse the firm is the larger the risk premium it adds to the price of the input and the higher its "effective" price. This causes substitution away from the uncertain-price input toward the certain-price input, resulting in a corresponding shift in the NLP. If the difference in the NLP is sufficiently large, it can cause a change in location, such as occurs when a increases from 0.05 to 0.10.

Similar results occur when we concentrate on a single firm (utility function), and vary the level of price uncertainty. For example, consider the firm with utility function parameter $a = 0.10$. We now examine the firm's optimal solutions as the degree of uncertainty in the price, $\tilde{p}_1$, increases from $\gamma = 0.2$ to $\gamma = 1.2$ (i.e., from $\tilde{p}_1 = 3.5 - 0.2\tilde{\varrho}$ where $\tilde{p}_1$ takes on values between 3.1 and 3.9 to $\tilde{p}_1 = 3.5 - 1.2\tilde{\varrho}$ where $\tilde{p}_1$ takes on values between 1.1 and 5.9). (Notice that when there is no price uncertainty, $\gamma = 0$, the same solution is optimal as for a risk-neutral firm.) When $\gamma = 0.2$: $z_1^* = 4.512$, $z_2^* = 3.546$, $x^* = 0$. When $\gamma = 0.8$: $z_1^* = 4.234$, $z_2^* = 3.779$, $x^* = 0$. When $\gamma = 1.2$: $z_1^* = 3.592$, $z_2^* = 4.454$, $x^* = s$. Thus, as the level of price uncertainty increases (i.e., as the spread of the distribution of the uncertain price increases), the firm substitutes the price-certain input for the price-uncertain input and the NLP adjusts so as to create greater net pull toward the source of the price-certain input. The underlying rationale is the same as for differences in risk aversion. For a given risk-averse firm, greater spread in the distribution of the price of the price-uncertain input around its mean creates greater perceived risk, which causes the firm to attribute a higher risk premium to the price of the price-uncertain input. This increases the effective cost of that input relative to the price-certain input, which creates a substitution and possible locational shift.

None of these results is counterintuitive and each suggests the possibility of developing theorems that will formally establish comparative statics (sensitivity analysis) results. Not all of the anticipated results have been formally established thus far; the more important ones to be developed are presented in the following sections.

Differences in Risk Preferences

We begin by considering the effects of differences in risk preferences. The pattern exhibited in the preceding example is formalized in the following theorem.

Theorem 5.5

In problem (DL5.1) [or (DL5.2)] let $\tilde{p}_1 = \bar{p}_1 - \tilde{\varrho}$ where $E(\tilde{\varrho}) = 0$ and $\tilde{p}_2 = p_2$ and $\tilde{p}_0 = p_0$ are deterministic. Let U_a and U_b be utility functions such that $r_b(\pi) \geqslant r_a(\pi)$ for all π.[3] Further, let (z_1^a, z_2^a, x^a) and (z_1^b, z_2^b, x^b) be optimal solutions to (DL5.1) [(DL5.2)] when $U = U_a$ and $U = U_b$, respectively. Then (a) $z_1^b \leqslant z_1^a$ and $z_2^b \geqslant z_2^a$; (b) $x^b \geqslant x^a$.

Since the approach (which is due to Martinich [1980] and Martinich and Hurter [1982]) used in the proof of theorem 5.5 is applicable to subsequent theorems, it will be presented here in detail. The proof will only be given for (DL5.1) [The proof for (DL5.2) is very similar but notationally more cumbersome.]

Proof (a): The proof is by contradiction; we assume that $z_1^b > z_1^a$ and show that (z_1^b, z_2^b, x^b) is not optimal for U_b, a contradiction. Let $v(d\varrho)$ be the probability measure of $\tilde{\varrho}$. Then the *expected profit* can be written as

$$\bar{\pi}^j = p_0 z_0 - (\bar{p}_1 + c_1(x))z_1^j - (p_2 + c_2(x))z_1^j;\ j = a, b \tag{5.30}$$

If ϱ is the realization of the random variable $\tilde{\varrho}$, then the realizations of profit are $\bar{\pi}^a + \varrho z_1^a$ and $\bar{\pi}^b + \varrho z_1^b$ for firms a and b, respectively. Without loss of generality, suppose $\bar{\pi}^a < \bar{\pi}^b$.[4] If $\bar{\pi}^a + \varrho z_1^a > \bar{\pi}^b + \varrho z_1^b$ for all ϱ such that $v(d\varrho) > 0$, then (z_1^b, z_2^b, x^b) cannot be optimal for firm b. Thus, there exists some realization of $\tilde{\varrho}$, say ϱ_0, such that $\bar{\pi}^b + \varrho z_1^b \lesseqgtr \bar{\pi}^a + \varrho z_1^a$ according as $\varrho \gtreqless \varrho_0$. Define $\pi_0 = \bar{\pi}^a + \varrho_0 z_1^a = \bar{\pi}^b + \varrho_0 z_1^b$. Since $\bar{\pi}^b > \bar{\pi}^a$, $\varrho_0 z_1^a > \varrho_0 z_1^b$. With $z_1^b > z_1^a$, by assumption, it follows that $\varrho_0 < 0$.

Von Neumann-Morgenstern utility functions are unique only up to a positive linear transformation. Thus, if U is a utility function, $\hat{U} = \alpha + \beta U$, $\beta > 0$, is also a utility function, and $\hat{U}$ and U generate identical preference *rankings*. U and $\hat{U}$ are said to be *strategically equivalent* and they have the same absolute risk-aversion functions, $r(\pi)$. Consequently, there is some utility function $\hat{U}_b$ which is strategically equivalent to U_b such that $\hat{U}_b(\pi_0) = U_a(\pi_0)$, $\hat{U}_b'(\pi_0) = U_a'(\pi_0)$ and $\hat{r}_b(\pi) = r_b(\pi) > r_a(\pi)$. In other words, the graph of $\hat{U}_b(\pi)$ is tangent to the graph of $U_a(\pi)$ at π_0 as shown in figure 5–2. There are several relationships between $\hat{U}_b(\pi)$ and $U_a(\pi)$ that should be noted: (a) $\hat{U}_b(\pi) < U_a(\pi)$ except at $\pi = \pi_0$, (b) $\hat{U}_b'(\pi) \gtreqless U_a'(\pi)$ when $\pi \lesseqgtr \pi_0$, and (c) $U_a - \hat{U}_b$ is strictly increasing moving away from π_0 in either direction.

For every realization, ϱ, of $\tilde{\varrho}$, $\bar{\pi}^b + \varrho z_1^b$ $(= \pi_0 + (\varrho - \varrho_0)z_1^b)$ is strictly farther away from π_0 than is $\bar{\pi}^a + \varrho z_1^a$ $(= \pi_0 + (\varrho - \varrho_0)z_1^a)$. (Recall that we have assumed $z_1^b > z_1^a$ and $\bar{\pi}^a < \bar{\pi}^b$).) Thus, for *every* ϱ

$$\begin{aligned} &U_a(\bar{\pi}^a + \varrho z_1^a) - \hat{U}_b(\bar{\pi}^a + \varrho z_1^a) \\ &< U_a(\bar{\pi}^b + \varrho z_1^b) - \hat{U}_b(\bar{\pi}^b + \varrho z_1^b) \end{aligned} \tag{5.31}$$

But

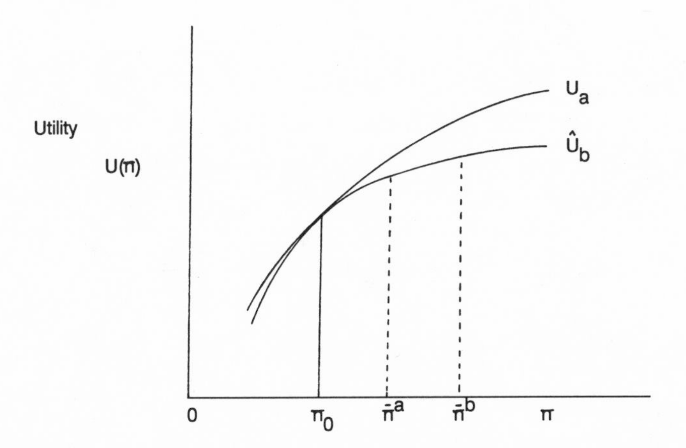

Figure 5–2. Relationship Between Utility Functions U_a and $\hat{U}_b$ in Proof of Theorem 5.5

$$E\hat{U}_b(\bar{\pi}^a + \tilde{\varrho} z_1^a) = EU_a(\bar{\pi}^a + \tilde{\varrho} z_1^a) - [EU_a(\bar{\pi}^b + \tilde{\varrho} z_1^b) - E\hat{U}_b(\bar{\pi}^b + \tilde{\varrho} z_1^b)] \quad (5.32)$$

and

$$E\hat{U}_b(\bar{\pi}^b + \tilde{\varrho} z_1^b) = EU_a(\bar{\pi}^b + \tilde{\varrho} z_1^b) - [EU_a(\bar{\pi}^a + \tilde{\varrho} z_1^a) - E\hat{U}_b(\bar{\pi}^a + \tilde{\varrho} z_1^a)] \quad (5.33)$$

The optimality of (z_1^a, z_2^a, x^a) for U_a insures that $EU_a(\bar{\pi}^a + \tilde{\varrho} z_1^a) > EU_a(\bar{\pi}^b + \tilde{\varrho} z_1^b)$. But (5.31) implies that the term in the [ ] on the right-hand side of (5.32) is strictly less than the term in [ ] on the right-hand side of (5.33); so $E\hat{U}_b(\bar{\pi}^a + \tilde{\varrho} z_1^a) > E\hat{U}_b(\bar{\pi}^b + \tilde{\varrho} z_1^b)$. But this contradicts the optimality of (z_1^b, z_2^b, x^b) for U_b. Consequently, the initial assumption that $z_1^b > z_1^a$ must be false so $z_1^b \leqslant z_1^a$. With z_0 fixed at $\bar{z}_0$, it also follows that $z_2^b \geqslant z_2^a$.

Proof (b): Again using contradiction, assume $x^a > x^b$. For any *given* input mix (values of z_1 and z_2), the optimal plant location for (DL5.1) also minimizes transport costs. Then, for (z_1^a, z_2^a):

$$c_1(x^a)z_1^a + c_2(x^a)z_2^a < c_1(x^b)z_1^a + c_2(x^b)z_2^a$$

or

$$\frac{z_1^a}{z_2^a} < \frac{c_2(x^a) - c_2(x^b)}{c_1(x^b) - c_1(x^a)}$$

When (z_1^b, z_2^b) is optimal,

$$\frac{z_1^b}{z_2^b} > \frac{c_2(x^a) - c_2(x^b)}{c_1(x^b) - c_1(x^a)}$$

Combining these inequalities and part (a) of this theorem, $z_1^a/z_2^a > z_1^b/z_2^b$ yields

$$\frac{c_2(x^a) - c_2(x^b)}{c_1(x^b) - c_1(x^a)} > \frac{z_1^a}{z_2^a} > \frac{z_1^b}{z_2^b} > \frac{c_2(x^a) - c_2(x^b)}{c_1(x^b) - c_1(x^a)} \tag{5.34}$$

which is a contradiction; we cannot have the results of part (a) and $x^a > x^b$. Therefore $x^a \leqslant x^b$. □

In many cases, both inputs must be used at a positive level if any output is to be produced (e.g., Cobb-Douglas production functions). Under these circumstances the inequalities in theorem 5.5(a) can be replaced with strict inequalities: $z_1^b < z_1^a$ and $z_2^b > z_2^a$.

Theorem 5.5 is consistent with our intuition that the more risk-averse firm will use less of the price-uncertain input to achieve a given level of output than a less risk-averse firm. Notice that theorem 5.5 does not apply to only risk-averse utility functions. This theorem holds for any pair of utility functions where the absolute risk-aversion function of one dominates the other. In addition, transport costs need not be concave in distance, nor must the production function satisfy strong conditions; this theorem holds for any transport-cost structure that is linear in quantity and for most substitutable production functions.

When more than one price is random the differential effects of risk preferences are difficult to characterize in general, but based on theorem 5.5, it is reasonable to conjecture that greater risk aversion would cause the firm to substitute away from the more risky of the inputs.

Changes in Output: Spatial and Production Invariance

In chapters 2–4 a substantial effort was expended to establish conditions under which the optimal ratio of input quantities z_1^*/z_2^*, and the optimal location x^*, are invariant with parametric changes in the level of output, $\bar{z}_0$. In the deterministic case, for the design/location problem, homotheticity of the production function was found to be an important condition leading to the *invariance* of the optimal input mix and location when output transport costs were nill. When output transport costs were included, degree-one homogeneity of the production function was usually required for such invariance. When input prices are uncertain but the firm is risk-neutral, it is straightforward to show that the same kind of invariance properties apply. In addition, it is easy to show that when only the output price is random, the same invariance properties hold as under certainty, regardless of the risk preferences of the firm. However, when *input* price uncertainty and general risk preferences are introduced, production homotheticity and homogeneity are insufficient to guarantee invariance, even when output transport costs are nill. This is demonstrated by the following example.

Example 5.4

Consider (DL5.1) where p_1 is random with distribution: $\text{Prob}(p = 1) = \text{Prob}(p_1 = 4) = 0.5$; and let $p_2 = 3$ be deterministic. Let $s = 1, c_1(x) = x$, and $c_2(x) = (1 - x)$. Suppose $F(z_1, z_2) = z_1 z_2$, $U(\pi) = -e^{-0.1\pi}$, and for simplicity let $p_0 = 0$.

(a) For $\bar{z}_0 = 1$ the optimal solution is $z_1^* = 1.20$, $z_2^* = 0.83$, $x^* = 0$, and $EU = -1.914$.

(b) For $\bar{z}_0 = 9$ the optimal solution is $z_1^* = 2.58$, $z_2^* = 3.49$, $x^* = 1$, and $EU = -7.558$.

Notice the change in both the optimal plant location and the optimal input mix. The underlying mechanics of this jump can be seen by studying the optimal solutions for various output levels at the two endpoints of the location space (we already know that the optimum location will be at some endpoint); these are given in table 5–2.

Under complete certainty (i.e., $\tilde{p}_1 = \bar{p}_1 = 2.5$), the optimal location for all output levels is $x^* = 0$, and the optimal input ratio $z_1^*/z_2^* = 1.60$. At

Table 5–2. Optimal Input Mixes and Locations for Example 5.4

	$x = 0$			$x = 1$		
$\bar{z}_0$	z_1^*	z_2^*	EU	z_1^*	z_2^*	EU
1	1.20	0.83	−1.914[a]	0.90	1.11	−1.930
3	2.02	1.48	−3.140[a]	1.53	1.96	−3.158
5	2.56	1.95	−4.451[a]	1.96	2.56	−4.454
7	2.99	2.34	−5.938	2.30	3.05	−5.909[a]
9	3.36	2.68	−7.640	2.58	3.49	−7.558[a]

[a] Global optimum.

low levels of output the amount of uncertainty, $\tilde{\varrho} z_1^*$, is small, as is the associated risk premium paid for input 1. As the level of output increases the amount of uncertainty, $\tilde{\varrho} z_1^*$, and the risk premium for input 1 increase[5] at both endpoints (but faster at $x = 0$). The firm uses proportionately less of the price-uncertain input and more of the price-certain input so the ratio z_1^*/z_2^* falls. Finally at $\bar{z}_0 \approx 5.1$ the risk premium is large enough so that input 2 becomes sufficiently "preferred" to input 1 to cause a jump in the optimal location to $x = 1$. Thus, even though the production function is homogeneous, the optimal input ratios and location are *not* invariant with respect to output when U is nonlinear and input prices are uncertain. In fact, even degree-one production homogeneity is not sufficient to guarantee invariance.

The preceding lack of invariance can best be seen by considering the expansion paths for the firm under various conditions. One of the properties of homothetic production functions is that isoquants for different output levels all have the same slope along any ray from the origin. When either input prices are known with certainty or the firm is risk-neutral, the

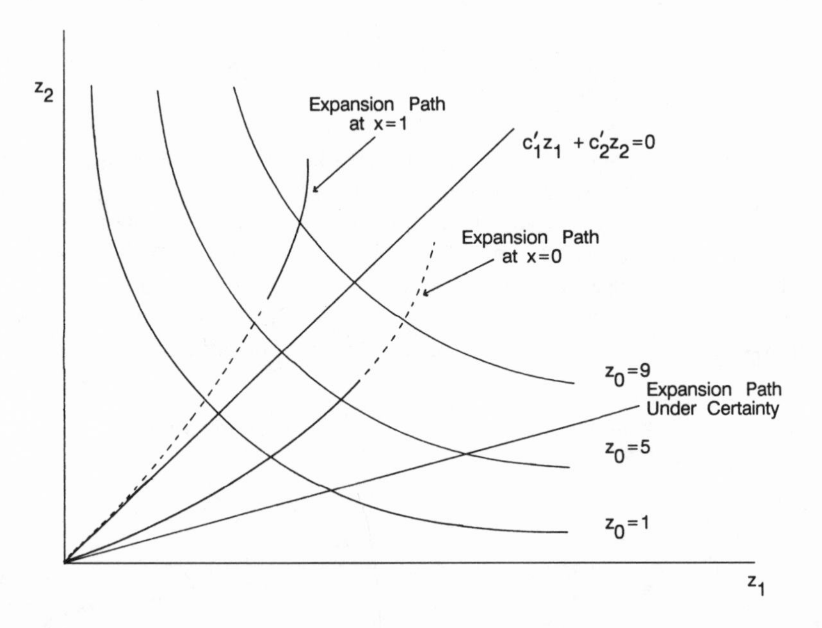

Figure 5–3. Expansion Path for Firm in Example 5.4

Firm is risk-averse, p_1 is uncertain, and F is homothetic. Solid lines indicate global expansion paths, dotted lines indicate expansion path at fixed endpoint that is not optimal. Jump in optimal location occurs at output level of approximately $\bar{z}_0 = 5.1$.

first-order conditions require equality of the (expected) delivered-price ratios with the slope of the relevant isoquant. The delivered prices are deterministic (or can be treated as such if U is risk-neutral) so their ratio could change only if the location changes. Consequently, the optimality conditions require the same optimal input ratios for all output levels, which also requires the same optimal location. That is, homothetic production functions generate expansion paths that are straight lines through the origin in input space. When input prices are uncertain and the firm is appropriately risk-averse, as the level of output changes the firm's expected profit changes; and thereby, the firm's level of risk aversion and its perceived risk premium for price-uncertain inputs change. Thus, the optimal input ratios change even at a fixed location because the effective delivered prices (actual price plus risk premium) change. The consequences of this output-dependent variation in delivered prices is illustrated in figure 5–3. Specifically, even with a homothetic (or homogeneous) production function, the expansion path is not linear and the same location need not be optimal for every output level.

Nevertheless, invariance *does* hold for (DL5.1) and (DL5.2) under degree-one production homogeneity when the firm's risk preferences satisfy a special, but common, condition. Specifically, invariance holds for those utility functions that exhibit constant proportional (relative) risk aversion (recall that $R(\pi) \equiv -\pi U''(\pi)/U'(\pi)$.) The assumption that $R'(\pi) = 0$ is a common one.

Theorem 5.6

Let some or all of the prices, $\tilde{p}_i$, be random variables and let $U(\pi)$ be a constant proportional (relative) risk-aversion utility function with $U'(\pi) > 0$, $U''(\pi) < 0$ for all π. Further, assume that the production function, F, is homogeneous of degree one. Then, in (DL5.1) and (DL5.2), the optimal input ratio, z_1^*/z_2^*, and the optimal plant location, x^*, are invariant with parametric changes in the rate of output, $\bar{z}_0$ (i.e., the expansion path is a straight line).

Proof: It is well known (see Pratt [1964]) that any utility function for which $R'(\pi) = 0$ is strategically equivalent to a function of one of the following forms: $U(\pi) = \ln \pi$; $U(\pi) = \pi^{1-c}$, $0 < c < 1$; or $U(\pi) = -\pi^{1-c}$, $c > 1$. The proof is constructed for (DL5.2) using $U(\pi) = \pi^{1-c}$; proofs for the other cases are identical. Let z_0^1 and z_0^2 be two arbitrary levels of output and let the ratio between them be $\alpha = z_0^2/z_0^1$. Let (z_1^1, z_2^1, x^1) and

(z_1^2, z_2^2, x^2) be optimal for z_0^1 and z_0^2, respectively. Define the random variable for profit, $\tilde{\pi}^j$ $(j = 1, 2)$ as

$$\tilde{\pi}^j = (\tilde{p}_0 - c_0(x^j))z_0^j - \sum_{i=1}^{2} (\tilde{p}_i + c_i(x^j)z_i^j) \tag{5.35}$$

$F(\alpha z_1^1, \alpha z_2^1) = \alpha z_0^1 = z_0^2$ and $F(z_1^2/\alpha, z_2^2/\alpha) = z_0^2/\alpha = z_0^1$ because F is homogeneous of degree one.

Since (z_1^1, z_2^1, x^1) is optimal for z_0^1,

$$EU(\tilde{\pi}^1) = E[(\tilde{\pi}^1)^{1-c}] \geqslant E[(\tilde{\pi}^2/\alpha)^{1-c}] = EU(\tilde{\pi}^2/\alpha) \tag{5.36}$$

For output level z_0^2, the expected utility of solution $(\alpha z_1^1, \alpha z_2^1, x^1)$ is

$$E[(\alpha\tilde{\pi}^1)^{1-c}] = E[\alpha^{1-c}(\tilde{\pi}^1)^{1-c}] = \alpha^{1-c}E[(\tilde{\pi}^1)^{1-c}] \tag{5.37}$$

The optimal expected utility for z_0^2 using (z_1^2, z_2^2, x^2) is

$$\begin{aligned} EU[(\tilde{\pi}^2)] &= E[(\tilde{\pi}^2)^{1-c}] = E[\alpha^{1-c}(\tilde{\pi}^2/\alpha)^{1-c}] \\ &= \alpha^{1-c}E[(\tilde{\pi}^2/\alpha)^{1-c}] \end{aligned} \tag{5.38}$$

From (5.36) it follows that (5.37) > (5.38) so $(\alpha z_1^1, \alpha z_2^1, x^1)$ must be optimal for output level z_0^2. □

We have shown that under the conditions of theorem 5.6 the expansion path for the firm is linear through the origin and the optimal location is independent of the level of output. Thus, under the stated conditions, the optimal solution for any arbitrary level of output can be used to find immediately the optimal solution for any other level of output. Although the conditions are special, they are not rare. This result holds no matter how many of the prices are random variables.

When only the output price is random, the invariance results of theorem 5.6 can be strengthened and generalized. Specifically, results identical to the deterministic case (theorem 2.8) hold, and they do not depend on whether or not the firm is risk-averse or risk-seeking.

Theorem 5.7

In (DL5.2) suppose only the output price is random and let F be homogeneous of degree k. Let (z_1^j, z_2^j, x^j) be the optimal solution for output level z_0^j, and let $z_0^2 > z_0^1$ be arbitrary output levels. Then x^2 $(\leqslant, =, \geqslant)$ x^1 and z_1^2/z_2^2 $(\geqslant, =, \leqslant)$ z_1^1/z_2^1 according as $k \lesseqgtr 1$.

The proof is essentially the same as for theorem 2.8 because for any fixed output level the firm's input mix and location decision reduce to a deterministic cost minimization problem. Thus a formal proof is omitted.

If the firm's production function is not homogeneous, the effects of changes in output level cannot be characterized in general. The firm's risk preferences, production structure, and transport costs all interact to determine the ultimate effect.

The inclusion of input-price uncertainty and general risk preferences not only can cause a loss of invariance, as illustrated by example 5.4, but can also lead to distortion of the firm's economies of scale and its *apparent* returns to scale in production. Recall that the "returns to scale" is a property of the production function, F. Formally, F exhibits (increasing, constant,decreasing) returns to scale if $F(\lambda z)$ $(>,=,<)$ $\lambda F(z)$, with $\lambda > 1$. That is, if when the quantities of all inputs (in a technologically efficient input combination) are doubled the resulting output is more than doubled, the production function is said to exhibit increasing returns to scale.

"Economies of scale" is a related concept based on the properties of the cost function. When average cost, $C(z_0)/z_0$ is (increasing,constant,decreasing) the firm is said to exhibit (decreasing,constant,increasing) economies of scale. Under constant (quantity-independent), deterministic input prices, (increasing,constant,decreasing) returns to scale in production generate (increasing,constant,decreasing) economies of scale. This equivalence makes it common to use cost-function information, which is often more easily measured empirically, to infer the form of firms' production functions. (An extensive duality theory of production and cost functions exists; see Shephard [1970] and Diewert [1974]. This duality forms the foundation of the theoretical work of EKR [1981].) Thus, if the cost function is convex, implying decreasing economies of scale, the inference is that the production function exhibits decreasing returns to scale.

If input prices are uncertain the deterministic cost function may be replaced by the expected cost function. An external observer will observe realizations for numerous values of z_0 and the actual costs associated with each. In an econometric sense, the observed cost function is an approximation of the expected cost function. The observer might then use the econometrically determined cost function to make inferences about the underlying production function without recognizing the uncertainty in input prices and the firms' risk preferences. Thus, if the observed cost function is (convex,linear,concave), then the normal conclusion is that the production function, F, exhibits (decreasing,constant,increasing) returns to scale. Such a conclusion may be incorrect.

With input-price uncertainty and nonneutral risk preferences the expected cost function may indicate diseconomies of scale even though the production function exhibits constant or increasing returns to scale. This result is related to our earlier conclusion that the expansion paths generated by a homothetic production function, which are straight lines under certainty, are nonlinear when input prices are uncertain and risk preferences are not neutral. For example, figure 5–4 illustrates the expected cost functions for a risk-averse firm under deterministic prices and under uncertain input prices, when F is homogeneous of degree one. In the former case the cost function is linear, exhibiting constant returns to scale and constant economies of scale. However, with input-price uncertainty, the expected cost function is convex, exhibiting decreasing economies of scale and *apparent* decreasing returns to scale.

A similar result obtains when, under certainty, the price of an input increases with the quantity purchased. Then a constant returns-to-scale production function can be associated with diseconomies of scale (i.e., a convex cost function). The uncertain-input-price case is very similar because the reason for the reduced economies of scale is that as the output (and thus input) levels increase, the disparity between the expected input

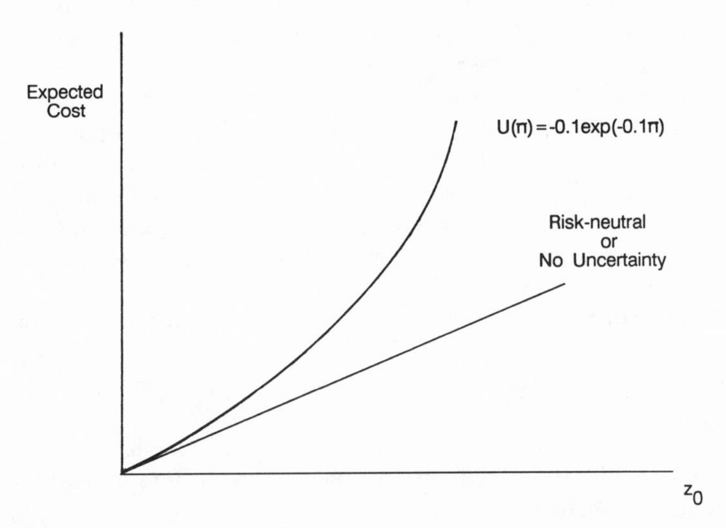

Figure 5–4. Expected Cost Function for Risk-Neutral Firm and for Risk-Averse Firm

F is degree-one homogeneous, p_1 is uncertain, and location is fixed. Expansion path under risk neutrality is the same as under no uncertainty.

price and the effective input price (i.e., adjusted for the risk premium) tends to increase, which creates an increasingly greater disparity between the firm's chosen input mix and its cost-minimizing mix.

It should be noted that the returns to scale bias is in the same direction for both risk-averse and risk-seeking firms. In both cases the price uncertainty and nonneutral risk preferences cause the firm to use input mixes that are not cost minimizing, thereby causing the actual average cost to be higher, and economies of scale lower (i.e., $C(z_0)$ is "bent" upward), than would be expected from the underlying production function. This bias also seems to be increasing with the level of risk aversion (absolute value $r(\pi)$) and the level of price uncertainty (γ) (see Martinich [1980, pp. 92–124] for an extensive discussion). When the parameters of the production function itself are treated as random variables, a similar disparity between the returns to scale and associated cost-function properties can result (see Hurter and Moses [1967]).

The immediately preceding discussion tacitly assumed a fixed plant location. Yet, when the expansion path is not a straight line (e.g., because of input-price uncertainty and risk aversion), we know that the optimal input mix changes with the rate of output. Consequently, the optimal plant location may change as well (although it will normally be at either $x^* = 0$ or $x^* = s$). When there is a shift in the optimal location, there will be a corresponding jump (discontinuity) in the associated expected cost function. For example, in (DL5.1) when p_1 is uncertain and U is appropriately risk-averse, the ratio z_1^*/z_2^* often decreases as $\bar{z}_0$ increases. If $x^* = 0$ at relatively low output levels, it is possible that $x^* = s$ (the location of the price-certain input) after $\bar{z}_0$ has reached some level, say $\hat{z}_0$. There will be an upward shift in the firm's expected cost function at $\hat{z}_0$ (this is illustrated in figure 5–5). At this point there is a discrete drop in the optimal quantity of the price-uncertain input, z_1^*, and its associated "risk" $\tilde{\varrho} z_1^*$. This reduction in risk offsets an increase in the expected cost associated with higher values of z_2^*. Consequently, the differences between the perceived returns to scale and actual returns to scale are accentuated by the possible discontinuity in the cost function that accompanies a shift in the optimal plant location.

The biases and discontinuities discussed above have important implications for empirical derivation of firms' production functions and their "apparent" decision making. Blair [1977] has shown how parametric uncertainty can lead to estimation errors in econometric research; yet the effects of parametric uncertainty and risk preferences on the "revealed" behavior of firms has not been considered extensively in the applied economics field.

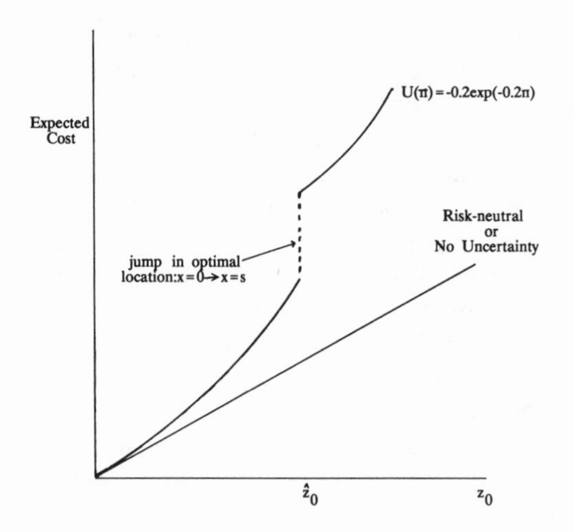

Figure 5–5. Expected Cost Function for Risk-Neutral Firm and for Risk-Averse Firm

F is degree-one homogeneous and p_1 is uncertain. "Jump" in optimal location for the risk-averse firm occurs at $z_0 = \hat{z}_0$ which causes a discontinuity in the expected cost function.

Changes in Price Uncertainty

As the level of price uncertainty (γ) increases in problems (DL5.1) and (DL5.2), intuition suggests that the risk-averse decision maker (firm) substitutes away from the input with the uncertain price, as illustrated by the results in table 5–1 for example 5.3. Example 5.3 further illustrates that changes in the level of price uncertainty can sometimes lead to a change in the optimal plant location. These patterns are surprisingly difficult to establish formally and in generality. We will begin our formal development by comparing the optimal solution under certainty to the case where the price of the first input is uncertain.

Theorem 5.8

Define (z_1^u, z_2^u, x^u) and (z_1^c, z_2^c, x^c) as the optimal solutions to (DL5.1) [or (DL5.2)] when input prices are $(\tilde{p}_1, p_2)$ and (p_1, p_2), respectively. As before, $\tilde{p}_1 = \bar{p}_1 - \gamma\tilde{\varrho}$ is random with $\bar{p}_1 = E(\tilde{p}_1)$, and p_2 is deterministic. (a) If U'' (<,>) 0, then z_1^u (≤,≥) z_1^c and z_2^u (≥,≤) z_2^c. (b) If U'' (<,>) 0, then x^u $(\geqslant,\leqslant)$ x^c. (c) If $U'' = 0$, then, for the two cases, the optimal solution is identical.

Proof: The proof will be given in terms of (DL5.1); the proof for (DL5.2) is essentially the same and is omitted for brevity.

(a) Suppose $U'' < 0$. It will be shown that $z_1^u > z_1^c$ leads to a contradiction. So assume $z_1^u > z_1^c$. Let

$$\bar{\pi}^j = p_0 z_0 - (\bar{p}_1 + c_1(x^j))z_1^j - (p_2 + c_2(x^j))z_2^j \tag{5.39}$$

for $j = u, c$; where $\bar{\pi}^j$ is the expected profit. Now $\tilde{\pi}^u = \bar{\pi}^u + \tilde{\varrho} z_1^u$. Since (z_1^u, z_2^u, x^u) is optimal for (DL5.1) under $(\tilde{p}_1, p_2)$,

$$EU(\tilde{\pi}^u) = EU[\bar{\pi}^u + \tilde{\varrho} z_1^u] > EU[\bar{\pi}^c + \tilde{\varrho} z_1^c] \tag{5.40}$$

For any realization of $\tilde{\varrho}$, since $z_1^u > z_1^c$, $|\varrho z_1^u| > |\varrho z_1^c|$ and then, in (5.40), theorem 2 of Rothschild and Stiglitz [1970] requires that $\bar{\pi}^u > \bar{\pi}^c$. However, (z_1^c, z_2^c, x^c) is optimal for $(\bar{p}_1, p_2)$ and, in that case, $U(\bar{\pi}^c) > U(\bar{\pi}^u)$. But $U' > 0$, so $\bar{\pi}^c > \bar{\pi}^u$, a contradiction. Consequently $z_1^u \leqslant z_1^c$ when $U'' < 0$. When $U'' > 0$ the same proof applies with the appropriate sign changes.

(b) The proof is almost the same as that employed earlier for part (b) of theorem 5.5 and is omitted.

(c) Risk-neutral utility functions are strategically equivalent to $U(\pi) = a + b\pi$. Substituting this into (DL5.1) gives first-order conditions identical to those for (DL2.1). □

As in the case of theorem 5.5, the weak inequalities in part (a) of theorem 5.8 can be replaced by strong inequalities whenever a positive rate of output requires $z_1^c > 0$ *and* $z_2^c > 0$.

Relative to the optimal solution under price certainty a risk-averse firm will substitute away from the price-uncertain input when uncertainty in the price of one input is introduced. Because this is a design/location problem and the design level of output is fixed, the risk-averse firm not only uses less of the price-uncertain input but more of the price-certain input when uncertainty is introduced. Consequently, the amount of the price-uncertain input that must be transported is reduced while the amount of price-certain input is increased. (The amount of output to be transported is fixed.) Thus, the net locational pull toward M_1 (i.e., $x = 0$) decreases and if the decrease is large enough, a change in the optimal location could occur (if $x^c = 0$).

General results are not available when incremental changes are made to an already uncertain input price or to a certain input price when other prices are uncertain. Results are available when location is fixed (see Stewart [1978]) and these show that as γ increases most risk-averse firms substitute away from the price-uncertain input. Generally, a risk-averse decision maker treats an increase in the level of price uncertainty for an input as similar to an increase in the expected price of that input. The adjustment to increased risk then not only includes a reduction in the use of the uncertain input, but also in expected profit. The decrease in expected profit may be interpreted as a kind of risk premium associated with the increasing levels of γ. It is important that formal results of this nature be developed for (DL5.1) and (DL5.2), that is, when location is a decision variable. As indicated earlier, no such general results, applicable to all distribution of $\tilde{p}_1$, are yet available. Results are available, however, for a special but common class of distributions.

Theorem 5.9

Consider (DL5.1) [(DL5.2)] with $U'' < 0$, with $\tilde{p}_1 = \bar{p}_1 - \gamma\tilde{\varrho}$, $\tilde{p}_2 = p_2$. Let (z_1^j, z_2^j, x^j) be an optimal solution for $\gamma = \gamma_j$, where, without loss of generality, $\gamma_2 > \gamma_1 > 0$. Let $\tilde{\varrho}$ have a distribution such that it can take on only one value less than zero (i.e., $\tilde{p}_1$ can take on only one value greater than its mean). Then (a) $z_1^2 \leqslant z_1^1$ and $z_2^2 \geqslant z_2^1$; (b) $x^2 \geqslant x^1$.

Proof: The lengthy and tedious proof is omitted here but is available in Martinich [1980]. □

The class of distributions to which theorem 5.9 applies includes all two-point distributions and many other common distributions such as three-point approximations of the Beta distribution where the distribution for $\tilde{p}_1$ is approximated by using a "pessimistic" price, p_1^a, an "optimistic" price, p_1^b, and a "most likely price", p_1^m. The distribution is skewed so that $E(\tilde{p}_1) > p_1^m$. These conditions are often reasonable approximations in many applications and may well reflect the way decision makers think about possible outcomes under uncertainty (i.e., their subjective probability distributions).

Example 5.3 suggests that results more general than theorem 5.9 probably can be established. For example, a natural class of distributions for $\tilde{p}_1$ that may lead to results similar to theorem 5.9 are price distributions that are symmetrical around the mean. More general results remain an area for future research.

Another area of possible research is investigation of the effects of changes in the expected value of a random input price. It would be reasonable to conjecture that the effects of changes in expected (average) price would be the same as the effects of price changes in deterministic models; namely, the firm would substitute away from inputs that have increases in their relative expected price, with a resulting effect on optimal location. This issue, however, has received only minimal attention and the validity of the preceding conjecture is not obvious.

Parametric Analysis: Production/Location Problems

Relatively precise parametric results could be obtained for (DL5.1) because, when the output level is fixed, an increase in one input level reduces the use of the second input and the resulting spatial effects are often clear. When output level is a variable, as in (PL5.1), parametric changes that affect the optimal output level can have ambiguous effects on input usage and location. Specifically, a parametric change that causes the optimal output level to increase may be associated with increased use of both inputs (not necessarily in the same proportion), or by increased use of one input and decreased use of the other; the resulting spatial consequences are then ambiguous. In spite of this additional complexity some unambiguous parametric results can be established. Throughout this section we will continue to assume that only the price of the first input is stochastic; all other parameters are deterministic.

Differences in Risk Preferences

Theorem 5.10

Consider (PL5.1) or (PL5.2) with prices $p_0, \tilde{p}_1$, and p_2, and let U_a and U_b be two utility functions such that $r_b(\pi) \geqslant r_a(\pi)$ for all π. Let $(z_0^a, z_1^a, z_2^a, x^a)$ and $(z_0^b, z_1^b, z_2^b, x^b)$ be optimal for (PL5.1) when $U = U_a$ and $U = U_b$, respectively. Then $z_1^b \leqslant z_1^a$.

Proof: The proof of the parallel theorem for (DL5.1), theorem 5.5(a), does not require nor utilize the fact that z_0 is fixed. Therefore, the proof of theorem 5.10 follows immediately. □

Note that general results regarding the relative values of z_0, z_2, and x have not been established. However, when the location, x, is fixed and (PL5.1) [or (PL5.2)] becomes a problem in the theory of production under uncertainty, some results are available. To some extent these results are analogous to those discussed above, and the results are stated here without proof (see Blair [1974]; Martinich [1980]).

1. If the production function, $F(z_1, z_2)$, exhibits decreasing marginal productivity (i.e., $F_{ii} \leqslant 0$) and if the marginal productivity of input i, $F_i(z_i, z_j)$, is increasing with input j (i.e., $F_{ij} \geqslant 0$) for $i = 1, 2$ and $j = 1, 2$, then $z_0^b \leqslant z_0^a$ in the problem described in theorem 5.10 with x fixed. Notice that the conditions on the production function are those most commonly assumed in economic analysis, and are often specified as second-order optimization conditions.

2. In addition to these conditions on the production function, if the marginal revenue $(p_0(z_0) + z_0 p_0'(z_0))$ is nondecreasing in z_0, then $z_2^b \leqslant z_2^a$. These results indicate that the decision maker with the more risk-averse utility function, U_b, uses less of each input and produces less output than the decision maker with the less risk-averse utility function, U_a. Furthermore, because the more risk-averse decision maker substitutes away from the price-uncertain input, z_1, it seems reasonable to conjecture that $z_1^b/z_2^b \leqslant z_1^a/z_2^a$. In (PL5.1) it would follow that $x^b \geqslant x^a$ if location were included as a variable (in (PL5.2) the spatial effect would be ambiguous because the pull of the output is counter to the input effects). However, no formal proof of this has been established in the literature, as yet. Notice that the most common application of this case is when the output market is perfectly competitive; in which case, the marginal revenue is constant.

When the production function is homothetic, stronger statements can be made about the relative use of inputs and the optimal location when comparing the decisions of firms with different levels of risk aversion. Many engineering-design relationships (the underpinnings of theoretical production functions) and almost all empirically (i.e., econometrically) estimated production functions are either homothetic or else they can be "satisfactorily" approximated by homothetic functions.

Theorem 5.11

Consider (PL5.1) with prices p_0, $\tilde{p}_1$, and p_2; let U_a, U_n, and U_s be three utility functions such that $U_a''(\pi) < 0$, $U_n''(\pi) = 0$, and $U_s''(\pi) > 0$ for all π. Assume F is homothetic. Then the optimal input ratios and locations are related as: (a) $z_1^a/z_2^a \leqslant z_1^n/z_2^n \leqslant z_1^s/z_2^s$, and (b) $x^a \geqslant x^n \geqslant x^s$, where $(z_0^j, z_1^j, z_2^j, x^j)$ is optimal for U_j.

Proof: It is easy to show that the optimal solution to (PL5.1) where $U = U_n$ is identical to the optimal solution under certainty with $\tilde{p}_1 = \bar{p}_1$. Further, as developed in chapter 2, when F is homothetic the optimal input ratios and plant location under certainty are invariant with respect to output level. Consequently, the input ratio z_1^n/z_2^n and location x^n are optimal for *every* level of output when $U = U_n$.

From theorem 5.5—note that $r_a(\pi) > r_n(\pi) > r_s(\pi)$—we know that for output level z_0^a, $z_1^a/z_2^a \leqslant z_1^n/z_2^n$ and $x^a \geqslant x^n$. Similarly, for output level z_0^s, $z_1^n/z_2^n \leqslant z_1^s/z_2^s$, and $x^n \geqslant x^s$. The statements of theorem 5.11 then follow immediately. □

When output transport costs are included, the preceding result does not necessarily hold; it depends on whether or not the returns to scale reinforce the input substitution effects or counteract them. If the production function is homogeneous of degree $k \leqslant 1$ (the more common case), then the results in theorem 5.11 extend to (PL5.2); but if $k > 1$, then they may not. (Conclusions drawn with respect to (PL5.2) depend on the assumption that the output market is at M_2, the source of the price-certain input.)

Although differences among firms in their degree of risk aversion can result in different optimal solutions under the same economic conditions, there is at least one case where the optimal input mix and location do not vary with risk preferences.

Theorem 5.12

Consider (PL5.1) or (PL5.2) where all prices can be random, but output price does not vary with output level. Let U_a and U_b be two *constant* absolute risk aversion utility functions with $r_a(\pi) = a$ and $r_b(\pi) = b$ for all π, and let $(z_0^j, z_1^j, z_2^j, x^j)$ be an optimal solution for $U_j, j = a, b$. Then if F is homogeneous of degree one, $(\alpha z_0^b, \alpha z_1^b, \alpha z_2^b, x^b)$ is optimal for U_a where $\alpha = b/a$.

Proof: All utility functions for which $r(\pi)$ is a constant are strategically equivalent to $U(\pi) = -\exp(-c\pi)$ with $c > 0$, a constant. (Notice that $r(\pi) = c$.) Thus, theorem 5.12 need only be proved for $U_a(\pi) = -\exp(-a\pi)$ and $U_b(\pi) = -\exp(-b\pi)$.

Define $\tilde{\pi}^j = \tilde{p}_0 z_0^j - (\tilde{p}_1 + c_1(x^j))z_1^j - (\tilde{p}_2 + c_2(x^j))z_2^j$; that is, $\tilde{\pi}^j$ is the random profit resulting from using solution $(\mathbf{z}^j, x^j)$ where $\mathbf{z}^j = (z_0^j, z_1^j, z_2^j)$. Notice that if F is homogeneous of degree one, then the random profit resulting from the solution $(\mathbf{z}^j/\alpha, x^j)$ is $\tilde{\pi}^j/\alpha$. Now by definition the solution $(\mathbf{z}^a/\alpha, x^a)$ is not preferred to $(\mathbf{z}^b, x^b)$ under U_b; so

$$E[-\exp(-b\tilde{\pi}^b)] \geqslant E[-\exp(-b(\tilde{\pi}^a/\alpha))] \qquad (5.41)$$

The expected utility under U_a for $(\alpha z^b, x^b)$ is

$$E[-\exp(-a(\alpha\tilde{\pi}^b))] = E[-\exp(-b\tilde{\pi}^b)] \tag{5.42}$$

Similarly, the expected utility under U_a for (z^a, x^a) is

$$E[-\exp(-a\tilde{\pi}^a)] = E[-\exp(b\tilde{\pi}^a/\alpha)] \tag{5.43}$$

From (5.41), (5.42) $\geqslant$ (5.43). But (5.43) is the optimal expected utility under U_a. Thus, $(\alpha z^b, x^b)$ is optimal for U_a. □

Results established in chapters 2–4 showed that under certain conditions the optimal input ratios and plant location are independent of parametric changes in output. Theorem 5.12 is similar in the sense that it is an invariance property, but it portrays invariance with respect to parametric changes in risk-aversion for utility functions within a special class of utility functions. Although the assumptions of a homogeneous production function of degree one, constant absolute risk aversion, and exogenous output price are restrictive, they are not uncommon assumptions. Similar to output invariance results, when the conditions of theorem 5.12 apply, the results have important computational implications. If (PL5.1) [or (PL5.2)] is solved for one constant absolute risk-aversion utility function it is solved for all such utility functions: the input ratios and location are the same and the input and output levels change according to α.

Changes in Price Uncertainty

The effects of changes in the level of price uncertainty on the optimal solutions to (PL5.1) and (PL5.2) are especially difficult to develop. This is an area that requires more research to develop new methods of analysis. A few results, which follow from theorems already developed, are presented below. (Again recall that the output market is at the source of input 2.)

Theorem 5.13

Let $(z_0^u, z_1^u, z_2^u, x^u)$ be an optimal solution to (PL5.1) [or (PL5.2)] with prices $(p_0, \tilde{p}_1, p_2)$ and $(z_0^c, z_1^c, z_2^c, x^c)$ be optimal under prices $(p_0, \bar{p}_1, p_2)$. Then $z_1^u \lesseqgtr z_1^c$ according as $U'' \lesseqgtr 0$.

Proof: Proof follows directly from theorems 5.10 and 5.8(c). □

Theorem 5.14

In (PL5.1), assume F is homothetic. Then the optimal input ratios and plant locations have the relationships z_1^u/z_2^u $(\leqslant, =, \geqslant)$ z_1^c/z_2^c and x^u $(\geqslant, =, \leqslant)$ x^c

according as $U'' \gtreqless 0$. The input inequalities are strict if $0 < z_1^c/z_2^c < \infty$ and $0 < z_1^u/z_2^u < \infty$.

Proof: Proof follows directly from theorems 5.11 and 5.8(c). □

Theorem 5.15

Consider (PL5.1) [or (PL5.2)] with $U'' < 0$, $\tilde{p}_1 = \bar{p}_1 - \gamma\tilde{\varrho}$, $\tilde{p}_2 = p_2$. Let $(z_0^1, z_1^1, z_2^1, x^1)$ and $(z_0^2, z_1^2, z_2^2, x^2)$ be optimal solutions for $\gamma = \gamma_1$ and $\gamma = \gamma_2$, respectively, where $\gamma_2 > \gamma_1 > 0$. Let $\tilde{\varrho}$ have a distribution such that it can take on only one value less than zero. Then $z_1^2 \leqslant z_1^1$.

Proof: Proof is similar to that use for theorem 5.9. □

Although more general results have not been established for incremental changes in the uncertainty of input prices, Mathur [1983] and Mai [1984a] have established some results for incremental changes in the level of output price uncertainty, which are extended in the following theorem. (Mathur and Mai perform their analysis assuming the optimal plant location is an interior point of $[0, s]$. As shown earlier in this chapter, such an occurence is unlikely under normal conditions. However, the following results hold even when optimal locations are at endpoints. The only difference in results is that in the following theorem the directional effects are of a weak form rather than strong form.)

Theorem 5.16

Consider (PL5.2) with $\tilde{p}_0 = p_0 - \gamma\tilde{\varrho}$, $\gamma > 0$, and $E(\tilde{\varrho}) = 0$; p_i, $i = 1, 2$ deterministic. Let F be homogeneous of degree k and $U'' < 0$. Then (a) z_1^*/z_2^* is (nondecreasing,constant,nonincreasing) and x^* is (nonincreasing, constant,nondecreasing) in γ according as $k \lesseqgtr 1$.

Proof: The proof follows from the fact that the optimal output level is decreasing in γ if $U'' < 0$ (see Mathur [1983] or Mai [1984a]), that the optimal input mix at any location is constant with output, and that the relative pull of the output market is decreasing (increasing) with output according as $k \gtrless 1$. The details of the proof are omitted for brevity. □

Solution Methods

Solution of the models presented in this chapter is only slightly more difficult than for the models given in chapter 2. The primary difference is that the production "subproblem" that has to be solved at any location is now a stochastic program, which is somewhat more cumbersome than a deterministic nonlinear program. Except for this difference the comments and suggestions made in chapter 2 apply to these models as well. A more extensive discussion of solving stochastic production/location problems is reserved for the next chapter, where computational issues are more crucial.

Conclusions

The material presented in this chapter extends the analysis of chapter 2 to an uncertain environment, and it acts as a foundation for the models in the remainder of the book. It was shown that the endpoint optimality results established under price uncertainty are as strong as those established under certainty. Under technological or transport-cost uncertainty, interior points of the line segment may be optimal locations; however, the restrictive conditions under which interior locations may be optimal makes this an uncommon occurrence.

Although the models in this chapter are mathematically complex, surprisingly strong parametric results have been developed for the case with random input prices. These results can be summarized as follows: (i) the more risk-averse is the firm, the less of the price-uncertain input it uses and the farther from the source of this input it locates. (ii) For a risk-averse firm, increases in the level of price uncertainty cause the firm to substitute away from the price-uncertain input and locate farther from its source location. (iii) Under special, but not unusual conditions, the optimal input mix and plant location do not vary with changes in key parameters: specifically, output levels in (DL5.1) and (DL5.2) and absolute risk aversion in (PL5.1) and (PL5.2). These latter results have obvious computational implications.

A number of aspects of stochastic production/location problems require further research. First, there are almost no comparative statics results for models with technological or transport-rate uncertainty. Only a couple of papers have been published so far which incorporate these other forms of uncertainty into the production/location framework, and to a large extent they have focused on endpoint optimality properties. Second, additional work should be done using a two-phase decision structure, where the location (and capital investment) decision is made before random parameters are realized, and then the values of production (flow) variables are selected after the randomness is realized. This two-phase approach would appear to be especially appropriate for a general production/investment/location model. Given the evolution of research that considered price uncertainty, it is likely that these research gaps will begin being filled soon.

The models and parametric analysis results developed in this chapter (and in chapter 6) are potentially very important as tools for developing and evaluating public policy alternatives. Facility design and location decisions are influenced by public policy instruments such as taxes, subsidies, incentives, and regulation; and these instruments are often directly used to influence the decisions of firms. For example, a particular government may use these instruments to attract certain kinds of industries to its confines. The impact on the site selection is usually through the production variables; consequently, the integrated production/location models should provide a basis for evaluating alternative policy instruments. The evaluations are especially complex when there is parametric uncertainty and general risk preferences. In chapters 7 and 8, however, some policy instruments will be evaluated within a stochastic environment and contrasted with the classical results from deterministic and nonspatial models.

Notes

1. Blair [1974] referred to $\bar{p}_i(z_i)$ as the "certainty equivalent price." But it appears that he meant "expected price," because the former varies with the decision maker's risk preferences.

2. Jensen's inequality states that for a random variable $\tilde{u}$ and a function F: $E[F(\tilde{u})] \gtreqless F[E(\tilde{u})]$ according as $F'' \gtreqless 0$.

3. If $r_b(\pi) > r_a(\pi)$ for some π in every interval, then the inequalities in (a), below, hold strictly.

4. If either U''_a or U''_b are of constant sign, then it can be shown that $z_1^b > z_1^a$ implies $\bar{\pi}^a < \bar{\pi}^b$.

5. The increase in the risk premium is caused primarily by the fact that $u(\pi) = -e^{-\pi}$ has increasing proportional risk aversion.

6 STOCHASTIC PRODUCTION/ LOCATION MODELS ON NETWORKS OR A PLANE

This chapter extends the analysis of stochastic production/location problems presented in chapter 5 from the line as the location space to network and planar spaces. The inclusion of parametric uncertainty in these more general spaces makes algorithmic solution of these problems, and especially analysis and characterization of optimal solutions using comparative statics methods, extremely difficult. This probably explains why so little published work exists on these types of production/location models. Although little work has appeared for either type of space, the work on stochastic network production/location problems is more extensively developed than the work on planar models, so the bulk of this chapter will focus on network models. The material in this chapter naturally builds upon the certainty models in chapters 3 and 4, and many of the methods and results extend to the stochastic case, though sometimes in altered form.

While there has been little work on network or planar stochastic production/location models, there is a modest literature on classical location models under uncertainty. The first section of this chapter gives a very brief review of some of this literature. It is not intended to be an exhaustive survey; rather it gives a general background of the types of results that have been developed for the stochastic versions of the median, fixed-charge, and Weber-type problems discussed earlier. In the next section stochastic production/location models on a network are formulated and solution characterization results are derived. Many of the node optimality results developed in chapter 3 extend to the stochastic case without change, using essentially the same proofs. Comparative static results are included, along with a numerical illustration. Planar models are presented next. The published work in this area is minimal and the results are quite limited; however, we have included some of our recent, partially completed work for planar spaces along with a numerical illustration. In the final section of this chapter, possible solution methods are suggested for network and planar stochastic production/location models.

Classical Location Models under Uncertainty

Uncertainty has been incorporated into several papers dealing with the location of facilities on a network. Much of the earlier work has been discussed by Handler and Mirchandani [1979]. Two sources of uncertainty are usually considered in such models: uncertainty in travel times (distances) along arcs and uncertainty in the demand at nodes. For example, Mirchandani and Odoni [1979] extend the certainty-case definition of network medians to the case where travel times on network links are random variables with known, discrete, probability distributions.[1] They establish conditions under whch the nodal optimality theorems of Hakimi and of Levy can be extended to these stochastic networks. (Stronger results for tree-network location problems are included in Mirchandani and Oudjit [1980]). Mirchandani [1980] generalizes the problem to allow random demands as well, and Weaver and Church [1983] have developed computational procedures for solving these models.

A number of articles have considered uncertainty within discrete-set location problems. Balachandran and Jain [1976] investigate the problem of determining the optimal location of plants, and their respective production and distribution levels, in order to meet demand at a finite number of centers. The possible locations of plants are restricted to a finite set of sites (e.g., nodes of a network) and the demands are allowed to be random. The cost structure of operating a plant is dependent on its location and is assumed to be a piecewise linear function of the production level, though not necessarily concave or convex. The expected value of production and transportation costs is to be minimized. Franca and Luna [1982] suggest a Benders Decomposition approach to solving a slightly simplified form of the problem. The problem they consider is a mixture of the classical plant location problem and the stochastic transportation problem. With demand at each destination a random variable, the problem is to minimize the sum of expected holding and shortage costs, linear shipping costs, and fixed construction costs.

In the previous models the objective functions are essentially to minimize expected costs. Jucker and Carlson [1976] generalize the single-plant, stochastic fixed-charge problem by representing the firm's risk preferences using a "mean-variance" utility function. (In their model both the demand for and price of the product can be random.) Although mean-variance utility functions have been criticized for some of their properties (they exhibit increasing absolute risk aversion), this representation does allow some insight into the effects of risk-averse preferences on the problem. Hodder and Jucker [1985] extend this model by considering the case where prices in various markets are correlated via their response to a common random factor. The result is a mixed-integer quadratic programming problem. However, for a given integer solution, the resulting quadratic programming problem is amenable to a very simple solution procedure and, consequently, reasonably large problems can be solved using branch-and-bound techniques.

Several stochastic variations of planar location models have been proposed in the literature. One of the earliest papers to deal with uncertainty in a planar location problem is an unpublished research report by Hurter and Prawda [1970]. This paper concerns the location of a new warehouse from which trucks are sent to pick up and deliver merchandise to customers. The delivery firm does not know, in advance, the locations and demands of its customers for a particular day, but it can compute the probability that it will be required to make a delivery to a particular customer or set of customers on a given day. Using the Euclidean norm for distances, the expected value of transportation costs is to be minimized subject to a set of chance constraints. (These constraints require that the probability of satisfying the demand of the jth customer is at least some prespecified amount.) The chance constraints incorporate trade-offs between the costs of not satisfying a customer's requirement and the costs of selecting an excessive capacity level for the warehouse. While not pursued in this volume, chance-constrained formulations sometimes describe actual business decision processes very well, and this approach may be a fruitful way of modeling stochastic production/investment/location decisions. The chance-constrained approach has been used by others for locational analysis. For example, Seppala [1975] used a chance-constrained formulation to invesitgate a multifacility problem. Aly and White [1978] used both unconstrained and chance-constrained formulations to address the problem of locating one or more new facilities relative to a number of existing facilities when the locations of the existing facilities, the weights between new facilities, and the weights between new and existing facilities are random variables. The new facilities are to be located such that expected distance traveled is minimized, and the distance measure used is Euclidean.

Cooper [1974] considers a Weber (Euclidean space) location problem in which the destinations are random variables with given probability distributions. The firm is to choose the plant location that minimizes the sum of the expected values of distances from the plant to the customers. Cooper claims that his results are valid for a wide class of probability functions which include any probability density function likely to be of relevance in an application. These claims are pursued by Katz and Cooper [1974]. In the same vein, Katz and Cooper [1976] discuss the case where the destinations (demand points) in the plane are independent random variables with specified probability densities. The problem is to find the facility location that minimizes the expected sum of Euclidean distances between the plant and the demand points. Katz and Cooper establish upper bounds for the minimum sum of distances in terms of solutions to the corresponding deterministic problems and the first two moments of the probability distributions. Furthermore, they show that the optimal location need not be in the convex hull of the means of each random customer location, and a sufficient condition for the optimal location to be in the convex hull is given. Cooper [1978], following up his earlier work with Katz, addressed the questions, "If the exact location of the destinations to which I ship material or services is different from what I stated, how much in error might the optimal solution (costs) be?" and "If I can only state that one or more destinations will be somewhere in some small area rather than at a given point, how much will this affect the optimal solution?" The answers are couched in terms of bounds on the solution. Wesolowsky [1977] and Drezner [1979] have worked on the same problem, where destination locations are random variables, but they used rectilinear distances. Drezner established bounds on the optimal solution, similar to Cooper's, but used rectilinear distances.

Drezner and Wesolowsky [1981] modify the previous stochastic Weber-type problems by assuming the weights that summarize cost and volume parameters are known only probabilistically, and the objective is to find the probability that the facility will be optimally located at any given point or in any region. Juel [1980, 1981] has established bounds on the solution, and Laridon [1984] used duality to develop computational algorithms for similar problems.

Many other stochastic location problems have been addressed in the literature that do not directly relate to the production/location type problems considered in this book. For example, Berman and Rahnama [1985] consider the location/relocation decisions of mobile servers, where the locations of servers can be changed at a cost. Travel times on the network arcs are random variables, and the objective is to minimize the long-term expected average cost of travel, taking into account both response-time costs and relocation costs. Chiu and Larson [1985] consider a service system where n mobile servers are garaged at one facility. Service demands arrive over time as a homogeneous Poisson process but are located over the service region according to an arbitrary probability law. The objective is to locate the garage facility so that the average cost of response (which is a weighted sum of mean travel time to a randomly serviced demand and the cost of a lost demand) is minimized. Another rather different example is given by Carbone and Mehrez [1980] who extend the minimax-distance single-facility location problem (i.e., minimize the greatest distance from a facility to any of its customers) to situations where the locations of prospective demand points are random variables. Two types of decisions are analyzed for this setting under the assumption of independent and identical normal distributions with the same means: locating on the basis of an expected value criterion or adopting a wait-and-see policy.

In each of the preceding papers, and in this volume, the parametric uncertainty is treated as neutral; that is, the uncertainty is due to environmental uncertainty aimed at neither helping nor harming the target firm. A very important alternative form of uncertainty relevant to production/location decisions is competitive uncertainty. For example, Hurter and Lederer [1985] and Lederer and Hurter [1986] explore the concept and existence of the equilibrium established for two costlessly mobile firms to be located in a planar market region. Here the uncertainty with which the firm must deal is the (anticipated) actions of its competitor in market areas, quantities offered for sale, and prices. The decisions involve choices of both delivered price schedules and firm locations. Each firm faces a production function and each is allowed to locate in the plane and to set discriminatory prices. Any transport cost function that is continuous in the firm location variable may be used. Demand is inelastic and marginal costs are constant with output but may differ with location and between competitors. An interesting result is that each firm increases its profit by locating so as to decrease the total cost to *both* firms of serving the market. Furthermore, firms will never locate coincidentally if they have identical production and transport costs. This type of competitive (game-theoretic) model will not be considered here. Instead it is left as an area of future research and is discussed further in chapter 9.

This brief review of location problems under uncertainty is not meant to be all inclusive but rather illustrative of the kind of work that has been and is being done. Potentially, these models may form the basis for model variations and solution algorithms for production/location problems.

Stochastic Production/Location Problems on a Network

As the preceding section indicates, there are many stochastic network facility location models in the literature, but they all ignore input substitution. Likewise, as demonstrated in chapter 5, there are a number of production/location problems on a line considered under uncertainty, but very little work has been published dealing with production/location problems on a network, and only two papers, Hurter and Martinich [1984a] and Louveaux and Thisse [1985], deal with the stochastic case. These two papers have taken different views regarding the timing of decisions and occurence of uncertainty. Hurter and Martinich utilize the more usual convention of assuming that the quantities of output and inputs as well as the location are chosen ex ante (i.e., before the random parameters are realized), and the transport cost *functions* are known with certainty. The total transport cost may be determined, and the profit corresponding to any realization of the random variable prices is maximized with respect to location when total transport cost is minimized. The same is true of expected utility of profit since it is a monotonically increasing function of profit. This leads to strong nodal optimality results similar to those in chapter 3. Louveaux and Thisse use an alternative approach in which the location and output capacity are chosen ex ante, but the quantities of output actually supplied to customers are chosen ex post (after random *demands* are realized). It is no longer possible to determine the total transport cost before the realization of the random variable. Consequently, nodal optimality results can be established only with less generality. Both approaches will be discussed, but the Hurter and Martinich [1984a] paper will form the heart of this section so as to be consistent with the focus in chapter 5, and because it provides a more extensive coverage of models and results.

Model Formulations

The models considered in this section are an integration of the deterministic network models in chapter 3 and the features of parametric uncertainty and risk-preference functions presented in chapter 5. Except where otherwise noted, the same notation will be used here as was used in those chapters. Parametric uncertainty will be restricted to the prices of inputs and outputs initially; in the subsection on node optimality, other forms of uncertainty will be considered. The firm is considered to be in a planning mode, where the values of location, capacity, and input utilization all are to be determined, prior to the realization of the random variables, so as to maximize the firm's expected utility of profit. Although adjustments to the values of the production variables will probably be made in the future after the realization of the random variables, our assumption here is that all of the values for decision variables must be determined simultaneously. Explicit ex post adjustment of production variables will be discussed later.

We summarize some of the notation from earlier chapters used here.

z_i = rate of utilization of ith input

z_i^k = rate of utilization of input i at plant k

z_0 = rate of production of product

z_{0m} = amount of output shipped to customer m

z_0^k = rate of output at plant k

z_{0m}^k = amount of output shipped from plant k to customer m

$F(z_1, \ldots, z_n)$ = production function

η = (N, A) the network with N the set of nodes and A the set of arcs

$N_1, \ldots, N_n$ = the nodes of η at which input sources are located

$N_{n+1}, \ldots, N_s$ = nodes of η that are not sources of inputs

$\tilde{p}_i$ = price per unit of input i at its source ($\tilde{}$ indicates random variable)

$\tilde{p}_0$ or $\tilde{p}_0(z_0)$ = price of product at market

A_{ab} = arc connecting nodes N_a and N_b

δ_{ab} = the length of A_{ab}. A distance scale is established so that $x = 0$ corresponds to N_a and $x = \delta_{ab}$ corresponds to N_b

${}_{ab}\phi_i(x)$ = the cost of transporting one unit of input i from N_a to x on A_{ab}

${}_{ab}\phi_0(x)$ = the cost of transporting one unit of output from x on A_{ab} to N_a

$c_i(A_{ab}, x)$ = the lowest cost for shipping one unit of input i from its source N_i to $x \in A_{ab}$, where either $c_i(A_{ab}, x) = c_i(A_{ab}, 0) + {}_{ab}\phi_i(x)$ where the least-cost route is through N_a or $c_i(A_{ab}, x) = c_i(A_{ab}, \delta_{ab}) + {}_{ba}\phi_i(x)$ where the least-cost route is through N_b

$c_{0m}(A_{ab}, x)$ = the lowest cost for shipping one unit of output from plant location x on arc A_{ab} to customer market m

$U(\pi)$ = von Neumann-Morgenstern utility function

As we have done throughout the book, we assume that the production function F, is a quasiconcave differentiable function with $F_i \equiv \partial F/\partial z_i > 0$ for all i and $F(0) = 0$. The arcs on the network may represent alternative modes of transportation, and, as assumed in chapter 5, the firm's utility function is stated as a function of its profits. For brevity, explicit formulations will be given only for the production/location cases; design/location formulations are omitted, but their structure can be determined easily by the reader. A design/location version of the network problem, however, will be used for numerical illustration; and invariance results for the design/location cases are presented formally.

Single Plant Models. Keeping with our earlier convention, we begin with a model of the case where output transport costs can be ignored. This assumption is appropriate when the output transport costs are relatively small or independent of location or when they are borne by the purchaser without influence on his or her buying decision. The firm's problem is to choose its plant location, input levels, and (indirectly) output level so as to maximize its expected utility:

$$\text{(PL6.1)} \quad \begin{aligned} &\underset{z_i, x}{\text{maximize}} && E(U(\tilde{\pi})] && && (6.1) \\ &\text{subject to} && z_i \geqslant 0 && i = 0, 1, \ldots, n && (6.2) \\ & && x \in \eta && && (6.3) \end{aligned}$$

where

$$\tilde{\pi} = [\tilde{p}_0(F(z_1, \ldots, z_n))]F(z_1, \ldots, z_n) - \sum_{i=1}^{n} [\tilde{p}_i + c_i(A_{ab}, x)]z_i \qquad (6.4)$$

and in choosing a plant location, x, it is implied that the arc on which it lies, A_{ab}, is chosen as well. If the output level is fixed a priori at $z_0 = \bar{z}_0$, then (PL6.1) becomes the corresponding stochastic design/location problem; we will designate this problem as (DL6.1).

In those situations where the costs of transporting product (output) are significant and borne directly by the firm, product transportation cost can be incorporated into the formulation of the problem with a little additional notation. We assume the firm is to supply several output markets, say M of them, each with its own stochastic inverse demand (price) function $\tilde{p}_{0m}(z_{0m})$, where z_{0m} is the amount of output sent to market m. Let ${}_{ab}\phi_0(x)$ be the cost per unit to ship output from location $x \in A_{ab}$ to N_a and $c_{0m}(A_{ab}, x)$ be the lowest cost of shipping one unit of output from $x \in A_{ab}$ to market m (notice that $c_{0m}(A_{ab}, x)$ equals either $c_{0m}(A_{ab}, 0) + {}_{ab}\phi_0(x)$ or

$c_{0m}(A_{ab}, \delta_{ab}) + {}_{ba}\phi_0(\delta_{ab} - x)$. In this case the firm's problem can be written as

$$\text{(PL6.2)} \qquad \begin{aligned} &\underset{z_i, z_{0m}, x}{\text{maximize}} && E[U(\tilde{\pi})] \\ &\text{subject to} && \sum_{m=1}^{M} z_{0m} = F(z_1, \ldots, z_n) && (6.5) \\ &&& z_i \geq 0 \qquad i = 0, 1, \ldots, n && (6.2) \\ &&& x \in \eta && (6.3) \end{aligned}$$

where

$$\tilde{\pi} = \sum_{m=1}^{M} [\tilde{p}_{0m}(z_{0m}) - c_{0m}(A_{ab}, x)]z_{0m} - \sum_{i=1}^{n} [\tilde{p}_i + c_i(A_{ab}, x)]z_i \qquad (6.6)$$

and the selection of x is implied to involve a selection of A_{ab} as well. When the quantities z_{0m} are set a priori equal to some levels $\bar{z}_{0m}$, we have the stochastic design/location problem which we will designate (DL6.2).

Multiplant Models. Here we suppose a firm is to determine the optimal locations, input usage rates, and production rates for K plants which supply M markets. As mentioned in chapter 3, if output transport costs are nill and each plant has a similar production function, it is usually optimal to locate all the plants at the same location unless there are limits on the availability of some inputs. So multifacility models are only meaningful if there are positive transport costs for the outputs. Therefore, we will formulate a multifacility generalization only of (PL6.2).

Using our notation from chapter 3, let (A^k, x^k) designate the location of plant k on some arc A^k, z_{0m}^k be the amount of output produced at plant k and sent to market m, z_i^k be the amount of input i used at plant k—so $\Sigma_{m=1}^{M} z_{0m}^k = F(z_1^k, \ldots, z_n^k)$—and without loss of generality, let $\tilde{p}_{0m}(\Sigma_{k=1}^{K} z_{0m}^k) = \tilde{p}_{0m}$. The firm's problem is then to

$$\text{(PL6.3)} \qquad \begin{aligned} &\text{maximize} && E[U(\tilde{\pi})] \\ &\text{subject to} && \sum_{m=1}^{M} z_{0m}^k = F(z_1^k, \ldots, z_n^k) && \text{for } k = 1, \ldots, K \\ &&& z_{0m}^k \geq 0,\ z_i^k \geq 0 && \text{for all } m, i, k \\ &&& x^k \in \eta && \text{for all } k \end{aligned}$$

where

$$\tilde{\pi} = \sum_{k=1}^{K} \left[\sum_{m=1}^{M} (\tilde{p}_{0m} - c_{0m}(A^k, x^k)) z_{0m}^k - \sum_{i=1}^{n} (\tilde{p}_i + c_i(A^k, x^k)) z_i^k \right]$$

and the selection of x^k implies selection of an arc, A^k. If the plant output levels, $z_0^k = \Sigma_{m=1}^M z_{0m}^k$, are set in advance, then this reduces to the multifacility, stochastic design/location problem which will be designated (DL6.3). (A variation of this problem is the case where the total amount of output sent to market m is fixed at some level, D_m, but the quantities z_0^k are variables. The following results hold for this case as well.)

Node Optimality

Uncertainty in Prices. From both a theoretical and computational viewpoint, it is quite valuable to identify conditions under which optimal plant locations can be restricted to the set of nodes in the network. When parametric uncertainty occurs only in the prices of the inputs and/or output, it is possible to establish node optimality results as strong as those for the deterministic network models formulated in chapter 3, *regardless of the firm's risk preferences.* We begin with the following:

Theorem 6.1

In problems (PL6.1)–(PL6.3) and (DL6.1)–(DL6.3), if all transport cost functions ${}_{ab}\phi_i(x)$ and ${}_{ab}\phi_0(x)$ are concave in x on all arcs, then (if a finite optimum exists) there exists an optimal locational solution which only contains nodes of η.

Proof: Because there is no randomness in the transport cost rates, for *any* values of the production variables, including the optimal values, the firm's total transport cost is a nonrandom function of plant location identical to that for the deterministic case. Because utility is a monotonic increasing function of profit, it follows that for the optimal values of the production variables the firm's optimal plant location(s) are those that minimize transport cost. It then follows immediately from Wendell and Hurter [1973b] that there exists an optimal locational solution that contains nodes of the network. □

This theorem is quite general in the sense that it sets no restrictions on the properties of the utility function other than for the most elementary, i.e., $U'(\pi) > 0$. The conditions on the transport cost functions are the same as those used in chapter 3; that is, the cost functions can look like those in figures 3–2 to 3–4 (see chapter 3), with jumps or fixed costs allowed to occur at nodes. Although theorem 6.1 does not exclude the possibility of optimal plant locations occuring on the interior points of arcs, this possibility can be excluded under certain conditions.

We begin with problems (PL6.1) and (DL6.1). To simplify the notation, without loss of generality, we assume that the price received for output, while random, is independent of the quantity of product sold; that is $\tilde{p}_0(z_0) = \tilde{p}_0$. Furthermore, define $(z_1^*, \ldots, z_n^*, A_{ab}^*, x^*)$ to be an optimal solution to (PL6.1) where $x^* \in A_{ab}$. We then have the following:

Theorem 6.2

In problems (PL6.1) and (DL6.1), if all ${}_{ab}\phi_i$ are concave in x, then the optimal plant location can only be at a node of η.

Proof: The proof is very similar to that for theorem 3.2, and so we will only give details for those parts that differ (see Hurter and Martinich [1984] for a self-contained proof). Using the exact same argument as that used for theorem 3.2, we can restrict our attention to the case where the following property, introduced in chapter 3, holds for the transport cost functions.

Property A. For each input i the least-cost route from source node N_i to x is via N_a for *all* $x \in A_{ab}$ or via N_b for *all* $x \in A_{ab}$.

In addition, from property WH2 (see chapter 3) we can restrict our attention to the case where the ${}_{ab}\phi_i$ are linear on A_{ab}; that is ${}_{ab}\phi_i(x) = r_i x$ for every $i \in M^a$ and ${}_{ba}\phi_j(\delta_{ab} - x) = r_j(\delta_{ab} - x)$ for every $j \in M^b$, where $r_i, r_j > 0$ are constants, M^a is the set of inputs with their least cost route to $x \in A_{ab}$ via N_a, and M^b is the set of inputs with their least cost route to $x \in A_{ab}$ via N_b. This fact, along with property A means that all the transport cost functions, $c_i(a_{ab}, x)$, are either increasing or decreasing linear functions on A_{ab}.

At $x \in A_{ab}$ profit can be written as

$$\tilde{\pi} = \tilde{p}_0 F(z_1, \ldots, z_m) - \sum_{i \in M^a} (\tilde{p}_i + c_i(A_{ab}, x)) z_i - \sum_{i \in M^b} (\tilde{p}_i + c_i(A_{ab}, x)) z_i \qquad (6.7)$$

The Lagrangian for (PL6.1) is then

$$L = E[U(\tilde{\pi})] + \mu(\delta_{ab} - x)$$

and the Kuhn-Tucker necessary conditions are

$$\left.\begin{aligned} \partial L/\partial z_i &= E[U'(\tilde{\pi})(\tilde{p}_0 F_i - \tilde{p}_i \\ &\quad - c_i(A_{ab}, x^*))] \leqslant 0 \\ z_i^*(\partial L/\partial z_i) &= 0 \end{aligned}\right\} \quad i = 1, \ldots, n \qquad \begin{aligned}(6.8\text{a})\\(6.8\text{b})\end{aligned}$$

$$\partial L/\partial x = E[U'(\tilde{\pi})\Big(-\sum_{i\in M^a} r_i z_i^* + \sum_{i\in M^b} r_i z_i^*\Big) - \mu^* \leqslant 0 \qquad (6.9\text{a})$$

$$x^*(\partial L/\partial x) = 0 \qquad (6.9\text{b})$$

$$\partial L/\partial \mu = \delta_{ab} - x^* \geqslant 0 \qquad (6.10\text{a})$$

$$\mu^*(\partial L/\partial \mu) = 0 \qquad (6.10\text{b})$$

$$\mu^* \geqslant 0 \qquad (6.11)$$

where * denotes optimal values, and notice that ${}_{ab}\phi_i'(x) = r_i$ for $i \in M^a$ and ${}_{ba}\phi_i'(x) = -r_i$ for $i \in M^b$.

Now because $U'(\pi) > 0$, condition (6.9a) is essentially the same as (3.4a) for the deterministic case, and the same arguments can be used to demonstrate that the following must hold for an optimal location, x^*, on arc A_{ab}:

$$\text{If} \quad \sum_{i\in M^a} r_i z_i^* - \sum_{i\in M^b} r_i z_i^* \begin{cases} > 0 \Rightarrow x^* = 0 = N_a & (6.12\text{a}) \\ = 0 \Rightarrow x^* = [0, \delta_{ab}] & (6.12\text{b}) \\ < 0 \Rightarrow x^* = \delta_{ab} = N_b & (6.12\text{c}) \end{cases}$$

(Notice that because the uncertainty in (PL6.1) is restricted to input and output prices, no random variables appear in (6.12a–c). If there were uncertainty in transport cost and/or in the production function (with the transport cost of output not being ignored) then (6.12a–c) would not be deterministic and the argument presented below for nodal optimality would not hold without modification.)

To complete the proof it is now sufficient to show that condition (6.12b) cannot hold. This will be proved by contradiction. So assume (6.12b) holds. Then the firm's optimal expected utility can be written as

$$EU\Big[\tilde{p}_0 F(z_1^*, \ldots, z_n^*) - \sum_{i\in M^a} (\tilde{p}_i + c_i(A_{ab}, 0))z_i^* \\ - \sum_{i\in M^b} (\tilde{p}_i + c_i(A_{ab}, \delta_{ab}) + \delta_{ab})z_i^* - \Big(\sum_{i\in M^a} r_i z_i^* - \sum_{i\in M^b} r_i z_i^*\Big)x^*\Big]$$

$$= EU\left[\tilde{p}_0 F(z_1^*, \ldots, z_n^*) - \sum_{i \in M^a} (\tilde{p}_i + c_i(A_{ab}, \delta_{ab}))z_i^* - \sum_{i \in M^b} (\tilde{p}_i + c_i(A_{ab}, \delta_{ab}) + \delta_{ab})z_i^*\right] \tag{6.13}$$

which is independent of the location variable x. Under these conditions, any location on arc A_{ab} is optimal. For *every* location $\bar{x}_0 \in [0, \delta_{ab}]$ the optimal input mix $(z_1^*, \ldots, z_n^*)$ must satisfy the necessary conditions:

$$E[U'(\tilde{\pi})(\tilde{p}_0 F_i - \tilde{p}_i - c_i(A_{ab}, 0) - r_i\bar{x})] = 0 \qquad i \in M^a,\ z_i > 0 \tag{6.14a}$$

$$E[U'(\tilde{\pi})(\tilde{p}_0 F_j - \tilde{p}_j - c_j(A_{ab}, \delta_{ab}) - r_j(\delta_{ab} - \bar{x})] = 0 \qquad j \in M^b,\ z_j > 0 \tag{6.14b}$$

Combining (6.14a) and (6.14b) yields

$$\frac{E[U'(\tilde{\pi})(\tilde{p}_i + c_i(A_{ab}, 0) + r_i\bar{x})]}{E[U'(\tilde{\pi})(\tilde{p}_j + c_j(A_{ab}, \delta_{ab}) + r_j(\delta_{ab} - \bar{x}))]} = \frac{F_i}{F_j} \tag{6.15}$$

There must be at least one $i \in M^a$ with $z_i^* > 0$ and one $j \in M^b$ with $z_j^* > 0$, otherwise the optimal location is surely at N_a or N_b. The right-hand side of (6.15) is constant in $\bar{x}$ while the numerator of the left-hand side is increasing and the denominator decreasing in $\bar{x}$. But equality can hold at most at one x and not for *every* $\bar{x} \in (0, \delta_{ab})$, which is a contradiction. Thus, (6.12b) cannot hold and the theorem is proved. □

In developing the proof for theorem 6.2, two assumptions were used for simplicity and they can be replaced with less stringent assumptions with the same proof remaining applicable. (i) Although transport costs were taken to be linear in quantity shipped, they need only be concave and increasing in quantity shipped for any fixed distance. (ii) There may be several arcs between any two nodes on the network (although only one was implicitly assumed in the proof). The several arcs, each of which might represent a different mode of transport, can be replaced by one composite arc for which a concave least-cost transport function can be constructed.

As with theorem 6.1, theorem 6.2 holds under relatively general conditions. Transport costs can be linear or concave (they can even have jumps at nodes), the production function can exhibit increasing, decreasing, or constant returns to scale, and the firm can be risk-averse, risk-neutral, or risk-seeking.

A strong form of node optimality can be established for problems (PL6.2), (DL6.2), (PL6.3), and (DL6.3) as well.

Theorem 6.3

For problems (PL6.2), (DL6.2), (PL6.3), and (DL6.3), if all ${}_{ab}\phi_i$ and ${}_{ab}\phi_0$ are concave, then an interior location of an arc can be an optimal plant location only if conditions equivalent to those in theorem 3.3 are satisfied.

A special case of theorem 6.3 which holds commonly in practice is the following.

Corollary 6.1

If the output transport costs on each arc are *strictly concave* in distance, then the optimal plant location(s) for problems (PL6.2), (DL6.2), (PL6.3), and (DL6.3) can be *only* at nodes of the network η.

Other Forms of Uncertainty. The fact that the preceding results follow rather directly from the deterministic models is due to the assumed form of parametric uncertainty. In this section we discuss node optimality when uncertainty is introduced through randomness in transport rates and/or in technological parameters. At the current stage of research, it is possible only to determine the circumstances under which nonnodal optima are possible. Further exploration is necessary to categorize completely the conditions leading to interior optimal locations.

Because we are primarily interested in the robustness of the nodal optimality result, we will base our argument on a modified version of the single-plant, single-market problem, (PL6.2). Among the more important modifications are the following:

M1: The cost of transporting a unit of input i along any arc A_{ab}, ${}_{ab}\phi_i(x)$, is linear in x.

M2: Combining M1 with property A we have

$$c_i(A_{ab}, x) = c_i(A_{ab}, 0) + \tilde{r}_i x \qquad \text{for all } i \in M^a$$

$$c_i(A_{ab}, x) = c_i(A_{ab}, \delta_{ab}) + \tilde{r}_i(\delta_{ab} - x) \qquad \text{for all } i \in M^b$$

M3: For simplicity, we assume there is only one output market and the travel of the *product* to market is through node N_a, i.e.,

$$c_0(A_{ab}, x) = c_0(A_{ab}, 0) + \tilde{r}_0 x$$

M4: The price of product, p_0, is independent of z_0

M5: Let $z_0 = \tilde{e}F(z_1, z_2, \ldots, z_n)$ [refer to chapter 5].

With these modifications we can write (PL6.2) as

$$\text{(PL6.2}'\text{)} \qquad \begin{array}{lll} \text{maximize} & E[U(\tilde{\pi})] & \\ \text{subject to} & z_i \geqslant 0 \quad i = 0, 1, \ldots, n & (6.2) \\ & x \in \eta & (6.3) \end{array}$$

where

$$\tilde{\pi} = [p_0 - \tilde{c}_0(A_{ab}, x)]\, \tilde{e}F(z_1, \ldots, z_n) - \sum_{i \in M^a} [p_i + \tilde{c}_i(A_{ab}, x)]z_i - \sum_{i \in M^b} [p_i + \tilde{c}_i(A_{ab}, x)]z_i \qquad (6.16)$$

The Lagrangian for (PL6.2′) is

$$L = E[U(\tilde{\pi})] + \mu(\delta_{ab} - x)$$

and the first-order optimality conditions are

$$\left.\begin{aligned} \partial L/\partial z_i &= E\{U'(\tilde{\pi})[p_0 - \tilde{c}_0(A_{ab}, x)]\tilde{e}(\partial F/\partial z_i) \\ &\quad - p_i - \tilde{c}_i(A_{ab}, x)\} \leqslant 0 \\ &\qquad z_i^*(\partial L/\partial z_i) = 0 \end{aligned}\right\} \quad i = 1, \ldots, n \qquad \begin{array}{r}(6.17\text{a}) \\ (6.17\text{b})\end{array}$$

$$\partial L/\partial x = E\left\{U'(\tilde{\pi})[-\tilde{r}_0\tilde{e}F(z_1, \ldots, z_n)] - \sum_{i \in M^a} \tilde{r}_i z_i + \sum_{i \in M^b} \tilde{r}_i z_i\right\} - \mu \leqslant 0 \qquad (6.18)$$

and (6.9b), (6.10a), (6.10b) and (6.11)

In (6.18) we have indicated the parameters, namely $r_0, r_1, \ldots, r_n$, and e, that might be random variables under the scenario of this section.

We now employ an argument similar to that used in our earlier node optimality proofs to identify when an interior location (i.e., $0 < x^* < \delta_{ab}$) is optimal. In that case, condition (6.9b) requires that (6.18) be met as an equality, and conditions (6.10a) and (6.10b) together require that $\mu^* = 0$. But $U'(\pi) > 0$, so for (6.18) to equal zero we must have $\partial\tilde{\pi}/\partial x = 0$. Then rewriting (6.16) as

$$\begin{aligned} \tilde{\pi} = {} & [p_0 - \tilde{c}_0(A_{ab}, 0)]\tilde{e}F(z_1, \ldots, z_n) - \tilde{r}_0 x\tilde{e}F(z_1, \ldots, z_n) \\ & - \sum_{i \in M^a} [p_i + \tilde{c}_i(A_{ab}, 0)]z_i - \sum_{i \in M^a} \tilde{r}_i x z_i \\ & - \sum_{i \in M^b} [p_i + \tilde{c}_i(A_{ab}, \delta_{ab})]z_i - \sum_{i \in M^b} \tilde{r}_i z_i(\delta_{ab} - x) \end{aligned}$$

yields

$$\partial\tilde{\pi}/\partial x = -\tilde{r}_0\tilde{e}F(z_1, \ldots, z_n) - \sum_{i \in M^a} \tilde{r}_i z_i + \sum_{i \in M^b} \tilde{r}_i z_i = 0 \qquad (6.19)$$

Using (6.19), π and therefore $EU(\pi)$ are independent of x. When $\partial\pi/\partial x = 0$, as required by (6.19), any point on arc A_{ab} must be optimal. However, as shown in the preceding subsection, it is impossible for the first-order condition $\partial L/\partial z_i = 0$ (as given by (6.17a)) to hold at all points on the arc A_{ab}.

When any or all of the parameters $r_0, r_1, \ldots, r_n$, and e are random, then $\partial\pi/\partial x$ as given by (6.19) is itself random and there is no satisfactory interpretation of $\partial\pi/\partial x = 0$. The proofs of nodal optimality based on the impossibility of $\partial\pi/\partial x = 0$ do not apply. If the firm is risk-averse (i.e., $U'' < 0$) an optimal interior location is possible just as it was in the case of a linear location space as discussed in chapter 5. Once again, the possibility of a nonnodal optimum depends on whether or not the variance in profit depends on the plant location. Consequently, although relative to a line, a network is a substantially more complex location space, the arguments given in chapter 5 for the linear space apply here as well. Notice that the randomness of $\partial\pi/\partial x$ only establishes the possibility of a nonnodal optimum. Whether such a solution will hold in a particular case will depend upon the curvature of $U(\tilde{\pi})$, the nature of the probability distributions of the random variables, and the resulting distribution of $\tilde{\pi}$. More research is required to characterize the conditions under which only nodal optima exist when parameters such as $r_i (i = 0, 1, \ldots, n)$ and e are random variables.

Ex Post Adjustment Models. Louveaux and Thisse [1985] present an alternative to the preceding models, where it was assumed that all decisions were made before the random phenomena were realized. Louveaux and Thisse stress the proposition that a long delay occurs between the time a location decision is made and the time the first output is produced. Things change during this delay period and the market conditions that prevail when output is available are not known when the plant location decision is made. Because of this, they study the location decision in the context of a two-stage stochastic model using a network spatial configuration.

In the first stage, the firm chooses its location, production level, and input levels before knowing the exact demands. In the second stage, it observes the realization of the random variables representing the demands on the local markets and decides upon the distribution of its production. The previously chosen production level may or may not be sufficient to cover the realized demands.

There are demand functions at each market node of the network where the quantity demanded at each node is determined by the price. The price received for product at each node is a random variable. In the second stage, an overall production level (and, through the production function an input level) and a location are assumed (i.e., the solution from the first-stage problem sets production level and location as parameters for the second stage). The firm chooses the quantities of product it will send to each market in order to maximize its profit. Profit is the realized price at each market node multiplied by the quantity of product shipped to that market minus the cost of transporting that product from the already determined plant location to the market node. The shipments to each node are constrained so they do not exceed the market demand at the given price and so that the sum of all shipments does not exceed the amount produced. This second-stage problem is a deterministic problem.

In the first stage, the location and the output rate levels are determined with the aim of maximizing the expected utility of profit. This optimization also involves the choice of input mix where a production function relates the level of product to the inputs used. The transportation cost for moving inputs from their source nodes to the plant location is included in the delivered price of the inputs. For a particular set of market prices, the profit function is defined conditionally upon the solution of the second-stage problem corresponding to the realization of the demand pattern. Thus, the firm chooses its capacity, input mix, and location before and its distribution pattern after observing the realization of demand.

Louveaux and Thisse are able to show that if the utility function is linear (i.e., the firm is risk neutral), an optimal solution exists with a location contained in the set of nodes. Furthermore, if the transport cost rate for the output is strictly concave in distance, the optimal location must belong to the set of nodes of the network. The location maximizing expected profit is generally not the same as the optimal location to the deterministic model with expected demands replacing the random demand variables. When the firm is risk-averse, every point along the network, including points interior to arcs, can be unique optimal locations for the firm depending upon the subjective probability distribution. According to Louveaux and Thisse, in the latter case the firm spreads the risk inherent in the geographical dispersion of demand by locating at an interior point of the network.

Network Location: Parametric Analysis

The network structure of the preceding stochastic production/location models makes results concerning the spatial effects of parametric changes almost impossible to characterize. The best we can do in most cases is to demonstrate that parametric changes can alter the optimum, and to determine on a case-by-case basis the effect of such parametric changes. However, in a few cases, specifically those that have already been identified in chapter 5, it is possible to determine conditions under which the optimal plant location(s) do not change with parametric perturbations. In this subsection we will first summarize the existing invariance results, and then we will present a numerical example to illustrate the possible effects of parametric uncertainty and risk preferences.

Two primary invariance results hold for network problems; the first applies to the design/location case whereas the second applies to the production/location case. These results are direct spatial generalizations of theorems 5.6 and 5.12, respectively. The proofs are essentially identical to those for Theorems 5.6 and 5.12, and are omitted. Self-contained proofs of these results are available in Hurter and Martinich [1984a].

Theorem 6.4

Consider the design/location problems (DL6.1)–(DL6.3) with output level fixed at some value(s), $\bar{z}_0$ ($\bar{z}_{0m}$ or $\bar{z}_0^k$). Assume the production function, F, is homogeneous of degree one. Let the utility function, $U(\pi)$, have $U'(\pi) > 0$, $U''(\pi) < 0$, $R'(\pi) = 0$ for all π. Then the optimal input ratios and plant location are invariant with parametric changes in output level. In problems (DL6.2) and (DL6.3), all of the $\bar{z}_{0m}$ and $\bar{z}_0^k$ or the total demands, D_m, must change parametrically in the same proportion.

The assumptions used in theorem 6.4 are admittedly restrictive but they are not uncommon in economic modeling. With these assumptons, the theorem indicates that the optimal plant design (i.e., input mix) and location are invariant with parametric changes in the level of output for decision makers with constant proportional risk-aversion functions. More precisely, the optimal input levels change in the same proportion as the fixed output rate. In problems (DL6.2) and (DL6.3), all the output distribution variables z_{0m}, z_{0m}^k must change in the same proportion (e.g., by q) for the result to hold. Equivalently, the *ratios* (z_{0i}/z_{0j}), (z_{0i}^q/z_{0j}^r) must remain constant in (DL6.2) and (DL6.3), respectively, as the total output is varied.

Theorem 6.5

Consider problems (PL6.1)–(PL6.3) with $\tilde{p}_0(z_0) = \tilde{p}_0$ and the production function F, is homogeneous of degree one. Let U_a and U_b be two utility functions with $U'_a(\pi) > 0$, $U'_b(\pi) > 0$, $U''_a(\pi) < 0$, $U''_b(\pi) < 0$, $r_a(\pi) = a$, $r_b(\pi) = b$ for all π, and let $(z_0^b, z_1^b, \ldots, z_n^b, x^b)$ be an optimal solution for U_b. Then $(\alpha z_0^b, \alpha z_1^b, \alpha z_2^b, \ldots, \alpha z_n^b, x^b)$ is optimal for U_a, where $\alpha = b/a$. (For (PL6.2) and (PL6.3) the z_0 is replaced by z_{0m} and z_{0m}^k, respectively.)

The assumptions of a degree-one homogeneous production function, constant absolute risk-aversion utility functions, and output price independent of z_0 are severe but are commonly employed in economic studies. Under these assumptions, theorem 6.5 states that the same input ratios and location will be optimal for all decision makers. If the problem is solved for one constant absolute risk-aversion utility function it is solved for all such functions.

The proofs of theorems 5.6 and 5.12 are independent of the location space; thus, other than for minor notational changes the proofs of theorems 6.4 and 6.5 are identical to those earlier proofs. The inclusion of multiple facilities is conceptually no more difficult; it is only notationally more cumbersome. The extent to which these theorems hold for models with other forms of uncertainty (transport rates or a technological parameter, e) has not been formally established. However, it appears that the same proofs should be valid for models with these kinds of uncertainty as well.

Some parametric results regarding effects on the production variables carry over from chapter 5 to the network models presented here. However, rather than presenting them as theorems, we will illustrate them as part of the following example from Hurter and Martinich [1984a].

Example 6.1

Consider problem (DL6.1) using the network depicted in figure 6–1, and assuming the level of output is fixed at $z_0 = \bar{z}_0 = 4.0$. The firm produces a single output using two physically distinct inputs, A and B, each of which is available at two source locations. Input A is available at the nodes labeled 1 and 2 at prices $\tilde{p}_1$ and $\tilde{p}_2$ while input B is available at the nodes labeled 3 and 4 at prices $\tilde{p}_3$ and $\tilde{p}_4$. The lengths of the arcs are as shown on figure 6–1, for example $\delta_{42} = 3$. With one exception, the per unit input transport costs for each input on each arc are $0.10d$ where d is the distance transported on the arc. The exception is for input 3 (i.e., input B from source node 3) where the per unit transport costs are $0.15d$ on arc A_{32} or ${}_{32}\phi_3(x) = 0.15x$.

The price of the first input is $\tilde{p}_1 = 3.5 + \gamma\tilde{\varrho}$ where the random variable $\tilde{\varrho}$ takes on values of $-2, -1, 0, 1, 2$ with equal probability and $\gamma > 0$. The value of γ determines the extent of the price uncertainty while $E(\tilde{p}_1) =$

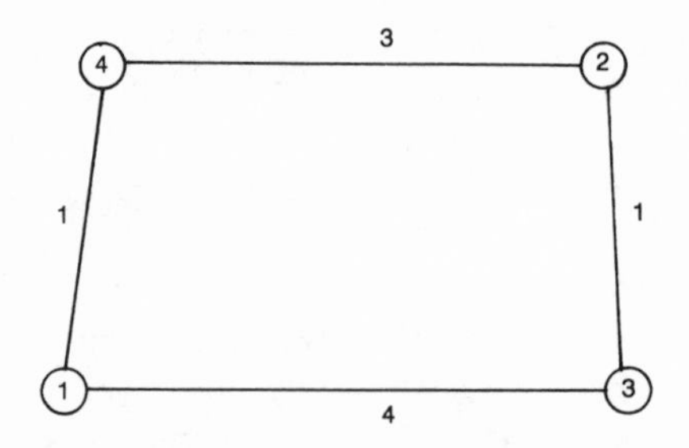

Figure 6–1. Network Location Space for Example 6.1

$\bar{p}_1 = 3.5$. The other prices are taken as deterministic for convenience and equal to $p_0 = 0, p_2 = 3.75, p_3 = 4.5, p_4 = 4.4$. The production function is $z_0 = (z_1 + z_2)^{0.5}(z_3 + z_4)^{0.5}$. The utility function is $U(\pi) = -\exp(-a\pi)$ where $U'(\pi) = a(\exp(-a\pi))$ and $U''(\pi) = -a^2(\exp(-a\pi)) < 0$ if $a > 0$; $r(\pi) = a$, $r'(\pi) = 0$; $R(\pi) = a\pi$, $R'(\pi) = a$. When $a = 0$, we define $U(\pi) = \pi$ and $r(\pi) = 0$, so that the utility function then is strategically equivalent to a risk-neutral utility function. This problem was solved for four values of a ($a = 0, 0.05, 0.10, 0.20$) and seven values of γ ($\gamma = 0, 0.2, \ldots, 1.2$). Tables 6–1, 6–2, and 6–3 give the best input mix and expected utility at each node for each value of a and γ, with global optima denoted in the tables. Notice that with no price uncertainty, that is when $\gamma = 0$, the optimal solution for all values of a is $x^* =$ node 1, $z_1^* = 4.535$, $z_2^* = z_3^* = 0$, $z_4^* = 3.528$.

Referring to tables 6–1 to 6–3, a form of parametric analysis can be performed. Consider a fixed level of uncertainty for $\tilde{p}_1$, say $\gamma = 1$. When $a = 0$ (i.e., a risk-neutral firm) the optimum is $z_1^* = 4.535$, $z_2^* = z_3^* = 0$, $z_4^* = 3.528$, and $x^* = N_1$. As the absolute level of risk aversion $r(\pi)$ is increased from 0 to 0.05 (i.e., $a = 0.05$) the optimal location remains at $x^* = N_1$, $z_2^* = z_3^*$ remain equal to zero, but z_1^* decreases to 4.290 and z_4^* increases to 3.730. In other words, as the absolute level of risk aversion increases from 0 to 0.05 the firm substitutes away from the price-uncertain input. With further increases in the risk-aversion parameter, a, to approximately $a = 0.07$, the risk premium associated with N_1, as a source for input A, increases until the firm begins to satisfy some of its needs for input A with purchases from node 2. The total quantity of input A falls and the total quantity of input B increases. By the time $a = 0.10$, z_1^* has decreased to 2.300, z_2^* has increased to 1.869, and z_4^* has increased to

3.838. These changes in input levels have caused the optimal location to change to $x^* = N_4$. As the value of a is increased still further, at approximately $a = 0.12$, the optimal values change dramatically to $z_1^* = 0$, $z_2^* = 4.324$, $z_3^* = 3.700$, $z_4^* = 0$, and $x^* = N_3$. This solution remains optimal for all values of $a > 0.12$.

Analogous results are shown in the tables where the absolute risk aversion $r(\pi) = a$ is constant and the level of price uncertainty in $\tilde{p}_1$, γ, is altered. For example, when $a = 0.10$ the optimal plant location is $x^* = N_1$ for $\gamma < 0.9$. As γ increases above this value, the optimal location switches to N_4, and then at still higher values of γ to $x^* = N_3$. The optimal input levels change accordingly.

As evident from this example, changes in problem parameters can cause changes in both the optimal location and production variable values. Because of the network structure it is not possible to describe in general how this optimal solution will change, especially if the network is complex and there are several random parameters. However, it is, in fact, true that if only one input price is random then, the more risk-averse the firm, the less of that input it should use ceteris paribus. The proof of this is the same as for the line segment space of chapter 5, and is consistent with the results given in tables 6–1, 6–2, 6–3.

Stochastic Production/Location Models on a Plane

The amount of published research on planar stochastic production/location models is miniscule. Katz [1984] considers the very special case of a location triangle (two inputs and one output) where the inputs are not substitutable and output price is random. Assuming the firm is risk-averse with decreasing absolute risk aversion, Katz's main conclusions are that with a decreasing returns-to-scale production function, the optimal plant location moves toward the output market and output level decreases as output price uncertainty increases or as mean output price decreases. Alperovich and Katz [1983] also use a location triangle, but they allow input substitution. They consider a planar *design/location* model where the output transport cost rate is a random variable. They show that if the firm exhibits increasing relative risk aversion (i.e., $R'(\pi) > 0$) and the production function is degree-one homogeneous, then the optimal plant location moves toward the output market as the output level increases. No exclusion theorems or other parametric results were derived.

Although there is a paucity of published work, a number of the results from chapters 4 and 5 can be extended easily to planar stochastic

Table 6–1. Solutions at Node 1 for Example 6–1[a]

a		γ					
		0.2	*0.4*	*0.6*	*0.8*	*1.0*	*1.2*
0.05	z_1	4.520	4.494	4.432	4.372	4.290	4.199
	z_2	0	0	0	0	0	0
	z_3	0	0	0	0	0	0
	z_4	3.540	3.560	3.610	3.660	3.730	3.810
	EU	−4.901[b]	−4.931[b]	− 4.980[b]	−5.048[b]	−5.132[b]	−5.232
0.10	z_1	4.507	4.444	4.348	4.233	3.410	2.330
	z_2	0	0	0	0	0.755	1.835
	z_3	0	0	0	0	0	0
	z_4	3.550	3.600	3.680	3.780	3.841	3.841
	EU	−24.12[b]	−24.70[b]	−25.66[b]	−26.96[b]	−28.47	−29.45
0.20	z_1	4.494	4.372	4.222	2.750	1.700	1.170
	z_2	0	0	0	1.415	2.465	2.995
	z_3	0	0	0	0	0	0
	z_4	3.560	3.660	3.790	3.841	3.841	3.841
	EU	−591.2[b]	−647.6[b]	−742.3	−848.0	−903.4	−934.3

[a] For $\gamma = 0$ or $a = 0$, optimal solution: $z_1 = 4.535$, $z_2 = 0$, $z_3 = 0$, $z_4 = 3.528$; this is a *global* optimum as well.
[b] A *global* optimum for these values of γ and a occurs at node 1.

Table 6–2. Solutions at Nodes 2 and 3 for Example 6.1 (for all values of γ and a)

		Node 2	*Node 3*
	z_1	0	0
	z_2	4.454	4.324
	z_3	3.592	3.700
	z_4	0	0
at $a = 0.05$	*EU*	−5.370	−5.300
at $a = 0.10$	*EU*	−28.30	−27.90[a]
at $a = 0.20$	*EU*	−780.0	−775.0[b]

[a] A *global* optimum for $a = 0.10$ and $\gamma = 1.2$ occurs at node 3.
[b] A *global* optimum for $a = 0.20$ and $\gamma = 0.8$, 1.0, or 1.2 occurs at node 3.

Table 6–3. Solutions at Node 4 for Example 6.1[a]

		γ					
a		*0.2*	*0.4*	*0.6*	*0.8*	*1.0*	*1.2*
0.05	z_1	4.408	4.384	4.336	4.267	4.188	4.103
	z_2	0	0	0	0	0	0
	z_3	0	0	0	0	0	0
	z_4	3.630	3.650	3.690	3.750	3.820	3.900
	EU	−4.923	−4.952	−4.999	−5.063	−5.144	−5.240[b]
0.10	z_1	4.396	4.336	4.255	3.640	2.300	1.590
	z_2	0	0	0	0.529	1.869	2.579
	z_3	0	0	0	0	0	0
	z_4	3.640	3.690	3.760	3.838	3.838	3.838
	EU	−24.33	−24.89	−25.81	−27.02	−27.82[b]	−28.27
0.20	z_1	4.384	4.267	3.340	1.820	1.150	0.790
	z_2	0	0	0.829	2.349	3.019	3.379
	z_3	0	0	0	0	0	0
	z_4	3.650	3.750	3.838	3.838	3.838	3.838
	EU	−601.1	−655.7	−741.7[b]	−791.3	−814.8	−827.8

[a] For $\gamma = 0$ or $a = 0$, optimal solution: $z_1 = 4.422$, $z_2 = 0$, $z_3 = 0$, $z_4 = 3.618$)
[b] A *global* optimum for these values of γ and a occurs at node 4.

production/location models. In addition, numerical illustrations can be helpful in gaining additional insight into the interactions inherent in such models and thereby helpful in generating conjectures regarding other possible solution properties. To demonstrate some of these results and interactions we will consider a stochastic generalization of problem (PL4.1); that is, we will assume a two-input production process and a location triangle as the location space (see figure 4–1 in chapter 4). Many of the results, however, apply to general n-input models.

Using the same notation as in chapters 4 and 5, the firm's problem is to

$$\text{(PL6.4)}\quad \begin{aligned} \underset{z_i, x, y}{\text{maximize}} \quad & EU(\tilde{\pi}) = [\tilde{p}_0 - r_0 d_0(x, y)]F(z_1, z_2) \\ & \qquad - \sum_{i=1}^{2} [\tilde{p}_i + r_i d_i(x, y)] z_i \qquad (6.20) \\ \text{subject to} \quad & z_i \geqslant 0 \qquad i = 1, 2 \\ & (x, y) \in R^2 \end{aligned}$$

Notice that we are assuming that the prices are random parameters. Under these conditions the exclusion theorem, theorem 4.1, extends immediately to this case.

Theorem 6.6

If $z_1^* > 0$ and $z_2^* > 0$, then an interior edge location cannot be optimal for (PL6.4).

The proof is essentially identical to that for theorem 4.1 and is omitted.

This theorem extends to a general n-input model as well; that is, the stochastic generalization of theorem 4.3 is valid as well and can be proven similarly.

In addition to these exclusion theorems, invariance theorems 5.6 and 5.12 apply to the planar case just as they do for the network models because the original proofs are independent of the location space. For brevity these are not stated explicitly, but it should be noted that these invariance results generalize to n-input and multifacility models using essentially the same proofs.

More general conjectures regarding the effects of parametric changes can be made on the basis of empirical evidence from numerical problems. We present the following example to motivate such conjectures.

Example 6.2

In (PL6.4) let $F(z_1, z_2) = (z_1 z_2)^{0.48}$. Let the price of input 1 at its source location, (2, 0), be $\tilde{p} = 3.5 + \gamma\tilde{\varrho}$ where Prob ($\varrho = k$) = 0.2 for $k = -2, -1, 0, 1, 2$ and $\gamma > 0$ is a parameter which controls the level of uncertainty; the price of input 2 at its source location, (0, 2), is 4.0, and the price of the output at its market location, (0, 0), is 10.0. Let the per unit transport cost of each input and output per unit distance be 0.20. Finally, assume the firm's utility function is $U(\pi) = -\exp(-a\pi)$, where $a > 0$ is a constant which controls the firms level of risk aversion (notice that $r(\pi) = a$). The optimal solutions for various values of the parameters γ and a are given below.

a	γ	z_0^*	z_1^*	z_2^*	x^*	y^*	z_1^*/z_2^*
0.05	1.0	6.271	6.661	6.880	0.46	0.49	0.968
0.10	1.0	4.153	4.261	4.558	0.45	0.50	0.935
0.20	1.0	2.710	2.680	2.975	0.42	0.50	0.901
0.20	0.7	4.427	4.564	4.861	0.44	0.50	0.939
0.20	0.5	6.848	7.329	7.511	0.47	0.49	0.976

As the firm's degree of risk-aversion, a, increases, it reduces output level and usage of the price-uncertain input (relative to the usage of the price-certain input). This results in a movement of the optimal plant location away from the output market and the source of the price-uncertain input and toward the source of the price-certain input. The same effects occur as the level of price uncertainty, γ, increases. Although no proof has been established to date, it is likely that these illustrative results hold more generally.

Solution Methods for Network and Planar Models

Network Models

The only published algorithmic work on network models has been for the one-plant models (PL6.1) and (PL6.2) (see Hurter and Martinich [1984a]). Consequently attention here is restricted to one-plant production/location problems; algorithms for multiplant production/location problems on a network with price uncertainty remain to be developed. Our earlier theorems usually permit us to limit our search for an optimal location for the one-plant production/location problem to the nodes of the network. A straightforward way of solving the problem then is to solve the (stochastic) production problem at each node—the optimal overall solution would then be the best of these nodal solutions. However, many problems of real interest will involve networks with many nodes. Under these conditions, it will be necessary or at least very desirable to reduce the number of nodes that are *explicitly* considered. Because many technological processes involve many inputs it also will be desirable to simplify the production problems to be solved at each node.

In many network location problems, bounding procedures can be developed to eliminate sets of nodes as candidates to be the optimal location. However, in production/location problems input substitution is permitted by the production functions and it becomes difficult to demonstrate that one node is a better location than another without at least partially solving the production problem at each node (i.e., obtaining individual bounds for each node). We begin this subsection by presenting the general procedure of the Hurter-Martinich algorithm. In presenting the algorithm, it is assumed that upper bounds, UB_i, are available for the stochastic production subproblem at each node, N_i. Following the algorithm, suggestions and details regarding possible bounds and other simplifications are presented.

Elimination by Bounding Algorithms.

Step 1. Select a node, N_a, and compute the optimal solution for its corresponding production problem, $\mathbf{z}^a$ (i.e., if the plant is located at N_a, the optimal solution is $\mathbf{z}^a = (z_{01}^a, \ldots, z_{0M}^a, z_1^a, \ldots, z_n^a)$). Store the (incumbent) optimal solution $\mathbf{z}^* = \mathbf{z}^a$, and the identity of the node $N^* = N_a$. Also set the lower bound $LB = EU[\tilde{\pi}(\mathbf{z}^a)]$ and set $\hat{N} = N - \{N_a\}$ ($\hat{N}$ is the set of nodes remaining to be searched).

Step 2. Select another node, N_i, from $\hat{N}$ and compute UB_i as defined below. If $UB_i \leqslant LB$ go to Step 3. If $UB_i > LB$, solve the stochastic production problem at node N_i to obtain solution $\mathbf{z}^i$. If $EU[\tilde{\pi}(\mathbf{z}^i)] \leqslant LB$ go to Step 3. If $EU[\tilde{\pi}(\mathbf{z}^i)] > LB$, set $LB = EU[\tilde{\pi}(\mathbf{z}^i)]$, $\mathbf{z}^* = \mathbf{z}^i$, and $N^* = N_i$.

Step 3. $\hat{N} = \hat{N} - \{N_i\}$. If $\hat{N} \neq \phi$ go to Step 2.

Step 4. Optimal solution is given by $\mathbf{z}^*, N^*, LB$.

This algorithm depends on the establishment of a "tight" upper bound, UB_i, for each node, N_i, *which is relatively easy to find.* One possibility is to let UB_i be the utility of the maximum expected *profit* given the plant is located at node N_i:

$$UB_i = U\left\{\begin{array}{lll} \underset{z_{0m},\, z_i}{\text{maximize}} & E(\tilde{\pi}) & \\ \text{subject to} & x = N_i & \text{and original problem} \\ & & \text{constraints} \end{array}\right\} \tag{6.21}$$

That is, we first maximize expected profit subject to the constraints of the original problem and $x = N_i$. This is equivalent to a deterministic problem with $p_i = E(\tilde{p}_i)$ and so UB_i is relatively easy to obtain. If the decision maker is risk-averse, as is commonly assumed, U is concave; it follows from Jensen's inequality that UB_i will be an upper bound with location fixed at N_i.

The degree to which UB_i is a tight upper bound depends on the curvature of the utility function as measured by U'' *and by* $r(\pi)$. To demonstrate this we use the following notation:

$\bar{\pi}^c$ = maximum expected profit at N_i
$\tilde{\pi}^*$ = The distribution of the profit associated with the expected utility maximizing solution
$\bar{\pi}^* = E(\tilde{\pi}^*)$

Note that $\bar{\pi}^c > \bar{\pi}^*$ and $UB_i = U(\bar{\pi}^c) > U(\bar{\pi}^*) > EU(\tilde{\pi}^*)$ which is the optimal solution to the original problem with $x = N_i$. The error in the bound can be written as the sum of two component parts:

$$UB_i - EU(\tilde{\pi}^*) = \{U(\bar{\pi}^c) - U(\bar{\pi}^*)\} + \{U(\bar{\pi}^*) - EU(\tilde{\pi}^*)\}$$

The difference expressed in the first component increases as $r(\pi)$ increases, while the difference expressed in the second component increases as $|U''|$ increases. Our limited computational experience indicates that UB_i is sufficiently tight, in many problems, to be useful. Tighter bounds can be developed but at the expense of increasing the amount and complexity of computation needed to determine them.

Simplification of Stochastic Production Problems. In many production/location problems, the number of physically distinct inputs that must be considered is relatively small (e.g., 2 to 10). However, when input is available from several different source nodes, the number of variables that must be considered production input variables rapidly increases, resulting in a relatively large, nonlinear, stochastic programming problem. This problem can sometimes be simplified by: (1) reducing the number of variables and (2) restricting the domain of some variables. The following theorem reflects some of these simplifications.

Theorem 6.7

In problems (PL6.1) and (PL6.2) assume U is concave. Let inputs $i_1, i_2, \ldots, i_r, \ldots, i_s$ be the same physical input available from s different source locations. The prices of the input at locations $i_1, \ldots, i_r$ are random variables, $\tilde{p}_i$, and the prices at $i_{r+1}, \ldots, i_s$ are deterministic at p_i. Define P_i to be the *delivered* price of the input to some plant location x; that is, $P_i = p_i + c_i(A_{ab}, x)$. P_i is either random or deterministic depending on p_i. Let z_i^* be the optimal rate of utilization of input i at location x.

(a) Denote the minimum of the deterministic delivered prices for input i as P_k (i.e., $P_k = \min_{\{i_{r+1}, \ldots, i_s\}} \{P_i\}$). Then $z_i^* = 0$ for all $i \in \{i_{r+1}, \ldots, i_s\} - \{i_k\}$.

(b) For $i = i_1, \ldots, i_r$ if $E(\tilde{P}_i) \geqslant P_k$, then $z_i^* = 0$

Proof: Part (a) states that in the absence of arc capacity or supply capacity constraints only the lowest deterministic delivered price supplier for any particular input need be considered. This requires no proof. To show part (b), let $E(\tilde{P}_i) > P_k$. In contradiction to the theorem, suppose $z_i^* > 0$. Then, by increasing z_k^* *by the amount* z_i^* while simultaneously reducing z_i^* to zero (i.e., replace the input from source location i with the same input from source location k) the expected profit will increase and the level of uncertainty in the resulting profit will decrease. Consequently, the new solution with $z_i^* = 0$ is preferred to the original and $z_i^* = 0$ is optimal. □

The usefulness of theorem 6.7 can be seen by considering example 6.1. Using this theorem at nodes 1 and 4, input 3 (i.e., physical input B with source node 3) can be dropped from the problem. When nodes 2 and 3 are considered, theorem 6.7 indicates that inputs 1 and 4 can be dropped and the resulting problem is deterministic. Usually there is no need to use more than a single source for a deterministically priced input, but, as the example indicates, it is possible that optimal usage requires several randomly priced suppliers of an input be used.

Lower Bounds and Ordering of Nodes. Tight lower bounds may be found by identifying a "good" node early in the sequential application of the algorithm. Since the upper bounds UB_i eventually will have to be computed for all nodes, it is convenient to compute upper bounds for all nodes at the beginning of the algorithm. The nodes then can be ordered and considered in decreasing order of their upper bounds. The production subproblem at the node with the greatest upper bound is solved completely and the value of the objective function is initially designated LB. The node with the second greatest upper bound is analyzed next and so on. In most cases, a strong lower bound will be found within a few iterations thus allowing rapid elimination of subsequent nodes.

When (6.21) is used as the upper bound, UB_i, a second approach for ordering nodes might be considered. In this second approach, the UB_i's again are computed at each node along with their corresponding solution, say $\mathbf{z}^{ic}$ (i.e., $UB_i = U\{E(\tilde{\pi}^i(\hat{\mathbf{z}}^{ic}))\}$). However, instead of using the UB_i's for ordering, the *expected utility* of each solution can be computed and these values used to order the nodes in decreasing order. If two nodes, N_a and N_b, have upper bounds UB_a and UB_b which are close in value, but the solution at N_a uses larger quantities of inputs with stochastic prices than the solution at N_b, then N_b is likely to be a better node than N_a even if $UB_b < UB_a$. The ordering scheme suggested is more likely to identify such cases.

Discussion of Algorithm. The algorithm solves very general, one-facility, stochastic production/location problems. In so doing, it requires only sightly more computational effort than would be required to solve deterministic or special stochastic location problems. The only additional computation is that involved in the solution of stochastic production problems at the relatively small number of nodes not eliminated by bounding.

When multiple facilities must be located, major computational difficulties arise in stochastic production/location problems. Under certainty or risk neutrality the (expected) cost-minimizing input mix can be determined at each node independently of the firm's activities at other nodes. After solving the production or input mix problem at each node to obtain production cost functions, plant location algorithms such as those discussed in chapter 3 can be used to solve the problem. These algorithms can also be used in stochastic problems if a mean-variance utility function is assumed for the firm (e.g., Jucker and Carlson [1976]). The key requirement in using these algorithms is that the objective function be additive and separable so that the profit (or utility) contributions of one facility do not affect the optimal input mix or marginal profit contributions of other facilities (except possibly through the constraints).

By way of contrast, under price uncertainty and nonneutral risk preferences, the optimal input mix at each facility depends upon the activities of the firm at its other facilities. The firm's activities at one facility help to determine its overall profit level and level of risk which, in turn, influence the risk premiums and effective input prices applicable to facilities at other nodes (see Hurter and Martinich [1984b] for a discussion and demonstration of how other plants affect the optimal solution for a given plant within a firm). One manifestation of this facility interrelationship is that even without facility capacity limitations more than one plant can share the same customers or market. For example, a portion of the demand of a particular market may be supplied by a facility using extensive amounts of a "stochastically priced" input. To supply all of the demand from this facility may add too much risk to the firm so the balance is supplied by a second facility which uses less of the price uncertain inputs.

Planar Models

Except that solution of the production subproblem at any planar location, (x, y), is more complicated, solution of stochastic planar models is identical to that for deterministic models. Specifically, the exact same methods discussed in chapter 4 can be applied to stochastic production/location problems on a plane, but whenever a production subproblem must be solved at some location (x, y) to obtain $z_i(x, y)$, a stochastic subproblem is solved. Global convergence cannot be proved, just as it could not be proved for the deterministic models. But our experiences have indicated that the algorithm does converge to an optimum and that the convergence (in number of iterations) is as rapid as for the deterministic models; actual computation time is somewhat slower because of the more complicated production subproblems.

Conclusions

The task of modeling network and planar production/location models under uncertainty clearly is not yet complete; nevertheless, the results already obtained do provide a foundation for future work. The inclusion of parametric uncertainty and risk preferences considerably enriches the theory of facility location and production. One of our major contentions is that integrated production/location models can be valuable tools for evaluating public policies. We have found the stochastic models to be especially valuable in identifying creative ways in which governmental entities can improve the operating environment for firms while satisfying their public goals and responsibilities more efficiently. In the next two chapters we illustrate how these models can be used and what conclusions they suggest regarding public policies.

Note

1. The network-median problem is the K-median problem, but with the location space being a network rather than a plane. See note 1 in chapter 3.

7 POLICY EVALUATION: TAXATION

Models of the sort developed in this volume have several important, potential, applications. On the one hand, they are normative in nature and may serve as the basis of planning models for individual decision-making entities. On the other hand, they may be descriptive of the general reaction of typical or "average" individual decision-making entities to exogeneous parametric changes. As descriptions of "average" decision-making entities (i.e., firms) and of their sensitivity to parametric manipulation, the models become potential instruments for the evaluation of governmental policies that influence parameters of firms' decision making. For purposes of discussion, three types of governmental activities will be considered here, namely: taxation, regulation, and locational incentives.

Taxes have been used by regional authorities for some time in an effort to persuade certain kinds of economic activity to locate in the region or to dissuade other forms of economic activity from locating there. For example, many states in the U.S. have maintained low corporate income tax rates in an attempt to attract industry and employment to a depressed region. On the other hand taxes are imposed to discourage the use of some inputs (e.g., gasoline) or input sources (e.g., tariffs).

Regulations are employed to restrict activities the regional authorities wish to curtail (e.g., rate of return on utility assets, air pollution, zoning). Regulation is the most direct (through not always the best) means available to regional planners as they attempt to develop their land-use plans. However, in many situations the use of direct regulation is restricted and recourse must be had to policy instruments, like taxes, that influence market parameters. Nevertheless, regulations play a very important role in helping to determine land-use patterns.

Locational incentives can take many forms, such as direct subsidy, tax abatement, or risk reduction. (Direct subsidy is similar to tax abatement. It distributes governmental resources—i.e., the taxes taken from individuals and corporations—to particular firms in the hope that the firms will locate a plant in the region.) Subsidies may take many forms including free or almost free land; construction of service facilities such as roads, water, or sewage specifically for the new plant; or providing free employee training. Tax abatements usually involve partial or complete elimination of property or other taxes for a specified number of years.

Risk reduction generally has taken the form of governmental loan guarantees which permit firms to borrow money at lower interest rates than would otherwise be the case. In a few instances, risk reduction has taken the form of agreements to purchase a stated number of units of product at a fixed, predetermined price—without regard to current market prices. A familiar example of this kind of risk reduction is the U.S. farm program. Another version of risk reduction that we feel is potentially very important is one where regional governmental units agree to guarantee the price of a price-uncertain input. This will be discussed at greater length in the next chapter.

Many of the potential governmental actions depend for their effectiveness on creating regional differences in taxes, subsidies, regulation, etc. However, one of the more interesting and potentially important results developed using production/location models in an uncertain environment as a policy evaluation tool is that even spatially *uniform* policy initiatives may favor location choices in one region over another. This is due to the inherent interrelationships among location, production rate, and input utilization choices, and makes the use of stochastic production/investment/location models a potentially valuable policy evaluation tool. In this and the next chapter we make a preliminary effort using the production/location models described in chapters 5 and 6. In this chapter we focus our attention on the effects of tax policy; in chapter 8 we consider locational incentives, risk reduction, and regulation.

The material in this chapter is based primarily upon the work of Martinich and Hurter [1985, 1987] and Hurter and Martinich [1987]. Two questions with respect to taxes are at the heart of the discussion. (1) Is the typical firm's optimal input mix, output rate, and plant location affected by the introduction of spatially uniform taxes, and if so, how? (2) Are some forms of business taxes more acceptable to firms and governments than are other types even though they are designed to raise the same revenue? These questions are addressed in an effort to shed some light on the form and mix of taxes likely to attract industry and to promote regional development.

We begin our investigation by incorporating various forms of business taxes into our deterministic production/location models. We find that in a deterministic environment the firm's optimal input ratios, plant location, and often output rate, are unaffected by the imposition of spatially uniform lump-sum, production, or income taxes. This explains why in most deterministic treatments of the spatial effect of taxes, the effect, if any, on the firm's facility decision stems from *regional differences in the level of taxation*; the form of taxation is not the center of analysis. Next, we incorporate the same taxes into the stochastic models of chapters 5 and 6, and demonstrate that except under special conditions these taxes do change the firm's optimal solution. More importantly, *the form of the tax makes a difference*. That is, two forms of taxation aimed at generating the same expected revenue can have dramatically different effects on the firm and be quite different in their desirability to the firm. These issues are discussed in a later section and illustrated with a numerical example. The primary issue identified is that by selecting (or altering) its form of taxation, a government may be able to generate additional revenue and accomplish its regional growth objectives while making the tax less objectionable to firms.

Taxation in a Deterministic Environment

To focus and simplify the discussion we will work with a generic form of a one-plant production/location model. To summarize our assumptions and notation, we consider a firm that produces a single product at the output rate z_0 using n inputs at the rates $z_1, \ldots, z_n$ according to a quasiconcave production function $z_0 = F(z_1, \ldots, z_n)$. The inputs are purchased from their source locations at given prices p_i while output is sold at a single market at a price p_0. Transport costs per unit for the output and inputs are $c_i(x)$, $i = 0, 1, \ldots, n$ where x is the unknown plant location. In this section all of the parameters are deterministic and so the problem for the firm is to:

(PL7.1)

$$\underset{z_1,\ldots,z_n,x}{\text{maximize}} \quad \pi = [p_0 - c_0(x)]F(z_1, \ldots, z_n) - \sum_{i=1}^{n} [p_i + c_i(x)]z_i \tag{7.1}$$

$$\text{subject to} \quad z_i \geqslant 0 \qquad i = 1, \ldots, n \tag{7.2}$$

$$x \in S \tag{7.3}$$

where S is the set of feasible sites; notice that S can be any of the location spaces mentioned in chapters 2–4. For comparative purposes, let $(z_0^*, z_1^*, \ldots, z_n^*, x^*)$ be the optimal solution to (PL7.1) which yields a profit $\pi^* > 0$.

Now suppose a taxing authority wants to raise an amount T in taxes from a firm within its region of jurisdiction (i.e., within region S). First consider a lump-sum tax of T on the firm. The lump-sum tax might be in the form of a license fee, a franchise tax, a permit, etc. We assume the lump-sum tax is applicable everywhere in S. With the lump-sum tax of T the objective of (PL7.1) becomes:

$$\underset{z_1, \ldots, z_n, x}{\text{maximize}} \quad \pi = [p_0 - c_0(x)]F(z_1, \ldots, z_n) - \sum_{i=1}^{n} [p_i + c_i(x)]z_i - T \tag{7.1a}$$

It is clear from inspection of the first-order conditions that the imposition of the lump-sum tax has no impact on the optimal values of $(z_0^*, z_1^*, \ldots, z_n^*, x^*)$ so long as $\pi^* - T > 0$.

Next consider a flat-rate income tax at the rate $t = T/\pi^*$. In this case the firm's objective function becomes:

$$\underset{z_1, \ldots, z_n, x}{\text{maximize}} \quad \pi = (1 - t)\left[(p_0 - c_0(x))F(z_1, \ldots, z_n) - \sum_{i=1}^{n} (p_i + c_i(x))z_i\right] \tag{7.1b}$$

Once again, the optimal values of the decision variables, $z_0^*, z_1^1, \ldots, z_n^*, x^*$ are unchanged and the after-tax profit is $(1 - t)\pi^* = \pi^* - t\pi^* = \pi^* - T$.

Suppose that an output or production tax of t^o dollars per unit of product is imposed (assume $t^o = T/z_0^o$ where z_0^o is the optimal output level under the tax, so T revenue will be raised). This tax is similar to that employed on the sale of gasoline or heating and cooking gas. In this case, the objective function for the firm becomes

$$\underset{z_1, \ldots, z_n, x}{\text{maximize}} \quad \pi = (p_0 - c_0(x) - t^o)F(z_1, \ldots, z_n) - \sum_{i=1}^{n} (p_i + c_i(x))z_i \tag{7.1c}$$

Assuming $z_i^* > 0$, $i = 1, \ldots, n$, the first-order conditions for (PL7.1) with (7.1) as the objective function are

$$(p_0 - c_0(x^*))F_i - p_i - c_i(x^*) = 0 \tag{7.4}$$

$$c_0'(x)F(z_1^*, \ldots, z_n^*) - \sum_{i=1}^{n} c_i'(x)z_i^* = 0 \tag{7.5}$$

where $F_i = \partial F/\partial z_i$. The first-order conditions when (7.1c) replaces (7.1) in (PL7.1) are

$$(p_0 - c_0(x^*) - t^o)F_i - p_i - c_i(x^*) = 0 \tag{7.4a}$$

and (7.5). A qualitative judgment on the impact of t^o can be made if it is assumed that x is fixed. Then $p_0 - c_0(x) - t^o < p_0 - c_0(x)$ meaning F_i in (7.4a) must be greater than F_i in (7.4). With diminishing marginal productivity the imposition of t^o reduces the optimal values of $z_0, z_1, \ldots, z_n$ and could alter the optimal location, x. However, if the production function is homothetic, as is often assumed, *and* if output transport costs are either insignificant or fixed (or demand is uniformly distributed in space), then the optimal input *ratios* and plant location will not be changed by the imposition of a ubiquitous output tax. (This result follows from theorem 2.6 and Hurter, Martinich, and Venta [1980], theorem 2). The reduction in optimal input levels (even with homothetic production functions, the quantity of each input used will be reduced although the ratios of input utilization remain unchanged) leads to a reduction in the level of output and in after-tax profits. The after-tax profits with an output tax, π^o, will be less than π^*, and so $\pi^o - t^o z_0^o = \pi^o - T < \pi^* - T$, and consequently an output tax is less desirable for the firm than lump-sum or flat-rate income taxes that raise the same revenue. (If output transport costs are significant, then the preceding results hold if F is degree-one homogeneous.)

The more common type of tax based on units sold is the sales tax. The sales tax is a percentage of the revenues generated by each unit of product sold. Let the sales tax be t^s. Then the objective of the firm whose revenues are subject to a sales tax is

$$\underset{z_1, \ldots, z_n, x}{\text{maximize}} \quad [p_0(1 - t^s) - c_0(x)]F(z_1, \ldots, z_n) - \sum_{i=1}^{n} (p_i + c_i(x))z_i \tag{7.1d}$$

In the formulation (7.1d), p_0 is the price paid by the purchaser at the market (which is assumed to be independent of z_0 and includes the sales tax), $p_0 t^s$ is the amount collected by government and $p_0(1 - t^s) - c_0(x)$ is the amount of revenue the firm receives after taxes and transportation costs. The imposition of a sales tax at the single-market place has an impact similar to that of the output tax since $p_0(1 - t^s) - c_0(x) = p_0 - c_0(x) - p_0 t^s$. Here, $p_0 t^s$ is an exogeneously determined constant and it plays a role in (7.1d) analogous to that played by t^o in (7.1c). Therefore the same conclusion about the impact of an output tax applies to the sales tax as well.

Finally we consider an input tax t^j per unit of input j. The value of t^j can be selected so that $t^j z_j^j = T$ where z_i^j is the optimal level of input i when there is a tax of t^j per unit on the use of input j. A ubiquitous input tax such as a property, utility, gasoline, or employment tax influences the firm's optimal input mix (if $z_j^* > 0$) and may alter the optimal location (even if the production function is homothetic), because the relative delivered prices of the inputs are altered. There is no question that $z_j^j < z_j^*$, and normally $z_0^j < z_0^*$ (this is assured if $\partial^2 F/\partial z_i \partial z_j \geqslant 0$), while use of the other inputs may increase or decrease. Thus, the input tax results in the firm using a less efficient input mix and, perhaps, a less efficient plant location which, together with the decrease in output, makes this form of tax especially undesirable to the firm. The optimal before-tax profit with the input tax π^j, is less than π^* and so $\pi^j - t^j z_j^j = \pi^j - T < \pi^* - T$.

The imposition of or change in spatially heterogeneous taxes (i.e., changes in spatial tax differentials) will influence the production and location decisions of firms. However, as shown above many common forms of business taxes are technologically and spatially neutral when they are imposed in a spatially uniform manner and in a deterministic environment. By technologically neutral we mean that the firm's optimal input utilization *ratios* are unaffected by the tax. By spatially neutral we mean that the imposition of the tax does not alter the optimal location for the firm. Regional authorities are interested not only in direct-tax revenues but also in the effect of the tax on regional employment and land use.

In a deterministic environment, the lump-sum tax and the flat-rate income tax are both spatially and technologically neutral, when employed in a spatially uniform manner. The output or sales taxes, with approximately homothetic technology (and no output transport costs), are also technologically and spatially neutral but they have a deleterious impact on output and input utilization. An input tax has the negative effect of reducing employment (and other input utilization) and output. It has the advantage of providing flexibility to the policy maker since it can be used to encourage or discourage the use of designated inputs and locations.

Taxation in an Uncertain Environment

In this section we analyze the impact of various forms of *spatially uniform* taxes in the stochastic environments discussed in chapters 5 and 6. Among other aspects of the impacts to be discussed is the technological and spatial neutrality of the taxes. We are particularly interested in identifying taxes that are spatially and technologically neutral in a deterministic environment but are not neutral in a stochastic environment.

As in chapters 5 and 6, input prices are taken to be random variables, and the firm is assumed to have general risk preferences represented by a von Neumann-Morgenstern utility function of profit, $U(\pi)$. We assume the firm is risk averse so $U'(\pi) > 0$ and $U''(\pi) < 0$. Our discussion is simplified by concentrating attention to the use of two (spatially distinct) inputs where the price of the first input at its source location is a random variable $\tilde{p}_1$, and the prices of the second input, p_2, and of output, p_0, are deterministic.

With these assumptions, the stochastic version of problem (PL7.1) is

$$\text{(PL7.2)} \qquad \begin{aligned} &\underset{z_1, z_2, x}{\text{maximize}} && EU(\tilde{\pi}) && (7.6) \\ &\text{subject to} && z_i \geqslant 0 \qquad i = 1, \ldots, n && (7.2) \\ & && x \in S && (7.3) \end{aligned}$$

where

$$\tilde{\pi} = [p_0 - c_0(x)]F(x_1, z_2) - [\tilde{p}_1 + c_1(x)]z_1 - [p_2 + c_2(x)]z_2. \qquad (7.7)$$

We have shown in chapter 5 that the introduction of uncertainty in p_1 coupled with risk aversion leads the firm to reduce its output level and usually to substitute input 2 for input 1. The extent of reduction in output and the extent of the substitution of one input for another increase with the degree of the firm's risk aversion and with the variability of $\tilde{p}_1$. The changes in output level and input mix will cause the firm's optimal location to adjust accordingly.

Our goal in the immediately following material is to obtain comparative statics or sensitivity analysis results for the variables z_0, z_1, z_2, and x with respect to the imposition of, or changes in, the level of various taxes. Unfortunately, we have found that mathematically precise results, in the sense of unambiguous directional changes, cannot be derived for all four variables simultaneously. One has to be fixed. An option common in the literature is to fix the output level so that (PL7.2) reduces to a design/location problem. With location restricted to a line (see chapter 5 for a description of the location space), the directional changes in input levels (i.e., the input ratio) and location, as a function of tax rates, can be determined. However, numerical experimentation with (PL7.2) in which z_0 is alllowed to vary indicates that the magnitude and direction of changes in input substitution and location with z_0 fixed are not accurate predictors of the changes in these variables when z_0 is allowed to vary.

Consequently, the alternative of fixing the plant location, x, in order to obtain unambiguous results for z_0, z_1, z_2 is employed here. The results, which apply to all locations, do allow meaningful, if not mathematically precise, inferences regarding the spatial impacts of taxes. These inferences are reinforced using numerical simulation. They appear applicable for more general location spaces such as R^2. Furthermore, the discussion of policy analysis follows more naturally with this approach, and it is more instructive for determining the spatial impacts of taxes.

Lump-Sum Taxes

Suppose that a regional government imposes a lump-sum tax of T on the firm, so that in (PL7.2) (7.7) is replaced by

$$\tilde{\pi} = [p_0 - c_0(x)]F(z_1, z_2) - [\tilde{p}_1 + c_1(x)]z_1 - [p_2 + c_2(x)]z_2 - T \tag{7.7a}$$

Then for any fixed location, x, the following holds.

Theorem 7.1

In problem (PL7.2) with x fixed and with lump-sum tax, T: (a) dz_1^*/dT, dz_2^*/dT, and $dz_0^*/dT \lesseqgtr 0$ according as $r'(\pi) \lesseqgtr 0$; (b) the optimal input ratio z_1^*/z_2^* is invariant with T if and only if $r(\pi)$ is constant (i.e., $r'(\pi) = 0$) or the production function F is homogeneous of degree one.[1]

Proof (a): We assume $z_1^* > 0$ and $z_2^* > 0$. For simplicity in notation write $P_i = p_i + c_i(x)$ as the price per unit of input i delivered to the plant located at x. The first-order conditions for (PL7.2) are

$$\partial EU/\partial z_1 = E[U'(P_0F_1 - \tilde{P}_1)] = 0 \tag{7.8a}$$

$$\partial EU/\partial z_2 = E[U'(P_0F_2 - P_2)] = 0 \tag{7.8b}$$

where $F_i \equiv \partial F/\partial z_i$. Totally differentiating (7.8a) with respect to z_1, z_2, and T (and noting that $\partial\pi/\partial T = -1$) yields

$$E[U'P_0F_{11}]dz_1 + E[U''(P_0F_1 - \tilde{P}_1)^2dz_1 + E[U''(p_0F_1 - \tilde{P}_1)(P_0F_2 - P_2)]dz_2 + E[U'P_0F_{12}]dz_2 = E[U''(P_0F_1 - \tilde{P}_1)]dT \tag{7.9a}$$

The total differential of (7.8b) is

$$E[U'P_0F_{21}]dz_1 + E[U''(P_0F_2 - P_2)(P_0F_1 - \tilde{P}_1)]dz_1 + E[U'P_0F_{22}]dz_2 + E[U''(P_0F_2 - P_2)^2]dz_2 = E[U''(P_0F_2 - P_2)]dT \qquad (7.9b)$$

where $F_{ij} \equiv \partial^2 F/\partial z_i \partial z_j$. Now define $|D|$ as

$$|D| = \begin{vmatrix} E[U'P_0F_{11}] + E[U''(P_0F_1 - \tilde{P}_1)^2] & E[U''(P_0F_1 - \tilde{P}_1)(P_0F_2 - P_2) + E[U'P_0F_{12}] \\ E[U'P_0F_{21}] + E[U''(P_0F_2 - P_2)(P_0F_1 - \tilde{P}_1) & E[U'P_0F_{22}] + E[U''(P_0F_2 - P_2)^2] \end{vmatrix}$$

Then Cramer's Rule yields

$$dz_1^*/dT = \frac{\begin{vmatrix} E[U''(P_0F_1 - \tilde{P}_1)] & E[U''(P_0F_1 - \tilde{P}_1)(P_0F_2 - P_2)] + E[U'P_0F_{12}] \\ E[U''(P_0F_2 - P_2)] & E[U'P_0F_{22}] + E[U''(P_0F_2 - P_2)^2 \end{vmatrix}}{|D|} \qquad (7.10)$$

We assume $|D| > 0$, which is equivalent to assuming that the second-order conditions for problem (PL7.2) are satisfied. Notice that first-order condition (7.8b) requires that $(P_0F_2 - P_2) = 0$. Consequently (7.10) reduces to

$$dz_1^*/dT = \frac{E[U''(P_0F_1 - \tilde{P}_1)] \quad E[U'P_0F_{22}]}{|D|} \qquad (7.11)$$

We know that $U'(\pi) > 0$ and $F_{22} < 0$ and consequently the sign of dz_1^*/dT is opposite the sign of $E[U''(P_0F_1 - \tilde{P}_1)]$.

At this point we introduce some notation. Let $B(P_1) \equiv U'(P_0F_1 - P_1)$; that is, $B(P_1)$ is the value of $U'(P_0F_1 - \tilde{P}_1)$ when $\tilde{P}_1$ takes on the value P_1. Recall that $r(\pi) = -U''/U'$, so

$$U''(P_0F_1 - P_1) = -r(\pi)U'(P_0F_1 - P_1) = -r(\pi)B(P_1)$$

The value of P_1 for which $B(P_1) = 0$ is labeled P_1^0, so $B(P_1^0) = 0$. It can be shown, from (7.8a) that $P_1^0 > \bar{P}_1 \equiv E(\tilde{P}_1)$ and $B(P_1) \lesseqgtr 0$ according as $P_1 \gtreqless P_1^0$. To emphasize the dependence on the value taken on by the random variable $\tilde{P}_1$, we will write $r(P_1)$ rather than $r(\pi)$ so

$$U''(P_0F_1 - P_1) = -r(P_1)U'(P_0F_1 - P_1) = -r(P_1)B(P_1)$$

Notice that the derivative of r with respect to P_1, $r'(P_1)$, is opposite in algebraic sign to $r'(\pi)$ since π decreases as P_1 increases. Let $v(dP_1)$ represent the probability measure of P_1. Then

$$E[U''(P_0F_1 - \tilde{P}_1)] = \int_0^{P_1^0} - r(P_1)B(P_1)v(dP_1) + \int_{P_1^0}^{\infty} - r(P_1)B(P_1)v(dP_1) \qquad (7.12)$$

Consider the case where $r'(\pi) = 0$. Then $r(P_1)$ is a constant and the right-hand side of (7.12) reduces to $r(P_1)E(B)$. The first-order condition (7.8a) and the definition of $B(P_1)$ require that $E(B) = 0$, so (7.12) = 0.

If $r'(\pi) < 0$ then $r'(P_1) > 0$. In the first integral of (7.12) $B(p_1) > 0$ since $P_1 < P_1^0$. Then, since $r(P_1) > 0$ and increasing in P_1,

$$0 > -\int_0^{P_1^0} r(P_1)B(P_1)v(dP_1) > -r(P_1^0)\int_0^{P_1^0} B(p_1)v(dP_1) \qquad (7.13a)$$

In the second integral of (7.12), $B(P_1) < 0$ and

$$-\int_{P_1^0}^{\infty} r(P_1)B(P_1)v(dP_1) > -r(P_1^0)\int_{P_1^0}^{\infty} B(p_1)v(dP_1) > 0 \qquad (7.13b)$$

Now combining (7.13a) and (7.13b) yields

$$E[U''(P_0F_1 - \tilde{P}_1)] > -r(P_1^0)E[B(P_1)] = 0 \qquad (7.14)$$

Consequently $dz_1^*/dT < 0$ if $r'(\pi) < 0$.

Similar reasoning shows that $dz_1^*/dT > 0$ if $r'(\pi) > 0$.

Repeating the same procedure to determine the sign of dz_2^*/dT yields

$$dz_2^*/dT = \frac{-E[U''(P_0F_1 - \tilde{P}_1)]E[U'(P_0F_{21})]}{|D|} \qquad (7.15)$$

Since $U' > 0$, $|D| > 0$, and $F_{21} > 0$, $E[U'(P_0F_{21})] \geq 0$ and the sign of dz_2^*/dT is opposite to the sign of $E[U''(P_0F_1 - \tilde{P}_1)]$. Based on the arguments used to determine the sign of dz_1^*/dT, it then follows that $dz_2^*/dT \lesseqgtr 0$ according as $r'(\pi) \lesseqgtr 0$.

Proof (b): If $r'(\pi) = 0$, then from part (a) z_1^* and z_2^* are invariant with T so z_1^*/z_2^* is invariant with T. So suppose $r'(\pi) \neq 0$, and divide (7.11) by (7.15) yielding

$$(dz_1^*/dT)/(dz_2^*/dT) = -F_{22}/F_{21} \qquad (7.16)$$

The input ratio z_1^*/z_2^* is invariant if and only if it equals the slope of the expansion path given by $(dz_1^*/dT)/(dz_2^*/dT)$. In other words

$$-F_{22}/F_{21} = z_1^*/z_2^* \quad \text{or} \quad F_{21}z_1^* = -F_{22}z_2^* \qquad (7.17)$$

Recognizing that $F_{21} = F_{12}$ and integrating (7.17) with respect to z_2 gives

$$z_1F_1 + z_2F_2 = F + c \qquad (7.18)$$

where c is a constant. If z_1 or z_2 equal zero, F is defined equal to zero, so c

must equal zero. So $F_1 z_1 + F_2 z_2 = F$. But, by Euler's theorem this holds only for degree-one homogeneous functions. □

Theorem 7.1 states that in the usual situation with decreasing absolute risk aversion ($r'(\pi) < 0$) the imposition of a lump-sum tax in (PL7.2) leads to a reduction in the optimal level of output and in the rate of utilization of both inputs. Furthermore, unless the technology can be represented by a production function that is homogeneous of degree one, the optimal input ratio will change. Unlike the situation in a deterministic environment, the lump-sum tax in an uncertain environment is not technologically neutral. Since the relative spatial economic pulls of each input depend on the relative quantities of inputs used, the imposition of a lump-sum tax alters the relative magnitudes of the pulls of the inputs and consequently the optimal plant location may differ from the no-tax case. Consequently a lump-sum tax in the environment described by (PL7.2) is not spatially neutral, unless the firm exhibits constant absolute risk aversion or the production function is degree-one homogeneous.

To illustrate theorem 7.1, assume $r'(\pi) < 0$ and $F(z_1, z_2) = z_1^{\alpha} z_2^{\beta}$. From (7.16) we have $(dz_1^*/dT)/(dz_2^*/dT) = -F_{22}/F_{21} = [(1 - \beta)z_1^*]/(\alpha z_2^*]$. Then if F is degree-one homogeneous, $\alpha + \beta = 1$ and $-F_{22}/F_{21} = z_1^*/z_2^*$. If $\alpha + \beta < 1$, then $-F_{22}/F_{21} > z_1^*/z_2^*$ and input 2 is substituted for input 1 as T increases so z_1^*/z_2^* is decreasing in T. This is stated more generally as a corollary to theorem 7.1.

Corollary 7.1

(a) If $r'(\pi) < 0$ and F is homogeneous of degree $\lesseqgtr 1$, then z_1^*/z_2^* is (decreasing,constant,increasing) in T. (b) If $r'(\pi) > 0$ and F is homogeneous of degree $\lesseqgtr 1$, then z_1^*/z_2^* is (increasing,constant,decreasing) in T.[2]

Proof:

$$\begin{aligned} d(z_1^*/z_2^*)/dT &= [(dz_1^*/dT)z_2^* - (dz_2^*/dT)z_1^*]/(z_2^*)^2 \\ &= \{E(U')P_0/(z_2^*)^2|D|\}\{F_{22}z_2^* + F_{12}z_1^*\} \\ &\quad \times \{E[U''(P_0F_1 - \tilde{P}_1)]\} \end{aligned} \tag{7.19}$$

The first term in braces in (7.19) is positive; the second term in braces is $\lesseqgtr 0$ according as F is homogeneous of degree $\lesseqgtr 1$, and the third term in braces is $\gtreqless 0$ according as $r'(\pi)$ is $\lesseqgtr 0$. Corollary 7.1 follows immediately. □

Income Taxes

The loss of the neutrality of these taxes in an uncertain environment is an interesting consequence of the use of stochastic rather than deterministic models. Of even greater interest is the direction of the bias associated with the taxes. The contrast between the impact of lump-sum and flat-rate income taxes is a case in point. Suppose a government imposes a flat-rate income tax (with full loss offset); the effect on firms is summarized below.

Theorem 7.2

In problem (PL7.2) with flat-rate income tax of rate t, let x be fixed. If $\pi(P_1) \geqslant 0$ for all values of P_1 with positive measure, then $dz_i^*/dt \lesseqgtr 0$ according as $R'(\pi) \gtreqless 0$, for $i = 0, 1, 2$.

Proof: First note that $\pi(P_1) \geqslant 0$ means that no values taken on by the random delivered price, $\tilde{P}_1$, lead to a negative profit.[3] We will employ the notation introduced in the proof of theorem 7.1. The first-order optimality conditions for (PL7.2) are

$$\partial EU/\partial z_1 = (1-t)E[U'(P_0F_1 - \tilde{P}_1)] = 0 \tag{7.20a}$$

$$\partial EU/\partial z_2 = (1-t)E[U'(P_0F_2 - P_2)] = 0 \tag{7.20b}$$

Totally differentiating (7.20a) and (7.20b) with respect to z_1, z_2, and t, using Cramer's rule and simplifying, it follows that the sign of dz_1^*/dt is opposite that of $E[U''(P_0F_1 - \tilde{P}_1)(\tilde{\pi})]$. Adopting the definition of $B(P_1)$ and P_1^0 from the proof of theorem 7.1, we define $R(P_1) = R(\pi(P_1))$ as the proportional (relative) risk aversion when $\tilde{P}_1$ takes on the value P_1. The sign of $R'(P_1)$ is opposite that of $R'(\pi)$. Then, after substituting appropriately we have an expression analogous to (7.12):

$$E[U''(P_0F_1 - \tilde{P}_1)(\tilde{\pi})] = \int_0^{P_1^0} - R(P_1)B(P_1)\nu(dP_1) + \int_{P_1^0}^{\infty} - R(P_1)B(P_1)\nu(dP_1) \tag{7.21}$$

If $R'(\pi) = 0$ for all π, then $R(P_1)$ is constant and (7.21) reduces to $R(P_1)E[B(P_1)] = 0$, because $E(B) = 0$ from (7.20a); so $dz_1^*/dt = 0$.

If $R'(\pi) > 0$, then $R'(P_1) < 0$ and in the first integral of (7.21), $B(P_1) \geqslant 0$. If $\pi(P_1) \geqslant 0$, then $R(P_1)$ is positive and decreasing in P_1, so

$$-\int_0^{P_1^0} R(P_1)B(P_1)\nu(dP_1) < -R(P_1^1)\int_0^{P_1^0} B(P_1)\nu(dP_1) < 0 \qquad (7.22)$$

In the second integral of (7.21), $B(p_1) \leqslant 0$ and by similar reasoning $\pi(P_1) \geqslant 0$, so

$$0 < -\int_{P_1^0}^{\infty} R(P_1)B(P_1)\nu(dP_1) < -R(P_1^0)\int_{P_1^0}^{\infty} B(P_1)\nu(dP_1) < 0 \qquad (7.23)$$

(Notice that the strict inequalities in (7.22) and (7.23) are correct because if $\tilde{P}_1$ is not degenerate there must be a positive probability of $\tilde{P}_1$ taking on a value strictly less than P_1^0.) Combining (7.22) and (7.23) gives

$$E(U''(P_0F_1 - \tilde{P}_1)(\tilde{\pi})] < R(P_1^0)E[B(P_1)] = 0$$

because $E[B] = 0$. Thus, $dz_1^*/dt > 0$.

If $R'(\pi) < 0$, a similar argument leads to $dz_1^*/dt < 0$.

The same reasoning can be used to prove $dz_2^*/dt \gtreqless 0$ according as $R'(\pi) \gtreqless 0$. □

A key assumption in this proof is that $\pi(P_1) \geqslant 0$ for all realizations of $\tilde{P}_1$ of positive measure. Often, $\pi(\bar{P}_1)$ (where $\bar{P}_1 = E(\tilde{P}_1)$) is sufficiently positive so that moderate variations in $\tilde{P}_1$ around $\bar{P}_1$ do not make $\pi(P_1)$ negative. In fact, if the utility of a negative profit is very negative, then $\pi(P_1) > 0$ for all P_1 will be required (through selection of $(z_0^*, z_1^*, z_2^*, x^*)$) for a finite optimum to exist. However, the assumption of $\pi(P_1) \geqslant 0$ for all P_1 is violated in one important case: when the production function is homogeneous of degree one. In that case

$$\begin{aligned}\pi(P_1) &= P_0F(z_1^*, z_2^*) - P_1z_1^* - P_2z_2^* \\ &= P_0(F_1z_1^* + F_2z_2^*) - P_1z_1^* - P_2z_2^*\end{aligned}$$

by Euler's theorem,

$$= (P_0F_I - P_1)z_1^* + (P_0F_2 - P_2)z_2^* = (P_0F_1 - P_1)z_1^* \qquad (7.24)$$

The last equality in (7.24) makes use of the first-order condition (7.20b). But $(P_0F_1 - P_1) \lesseqgtr 0$ according as $P_1 \gtreqless P_1^0$. Then in the first integral of (7.21), $R(P_1) > 0$ regardless of the sign of R' and $B(P_1) \geqslant 0$ so the entire integral is negative.[4] Likewise, in the second integral of (7.21), $R(P_1) < 0$ regardless of the sign of R' and $B(P_1) \leqslant 0$, so the entire integral, and thus (7.21), is negative. It then follows that $dz_1^*/dt > 0$ regardless of the sign of R'.

In fact, when F is homogeneous of degree one, rather strong invariance results with respect to the income tax rate have been established by Martinich and Hurter [1987].

Theorem 7.3

Let $(\mathbf{z}^*, x^*) \equiv (z_1^*, \ldots, z_n^*, x^*)$ be an optimal solution for (PL7.2) when $t = 0$. If F is homogeneous of degree one, then for any $0 < t < 1$, $(\mathbf{z}^*/(1 - t), x^*)$ is optimal for (PL7.2), and $U^*(t) = U^*(0)$ where $U^*(t)$ is the optimal expected utility under income tax rate t.

Proof: Because F is homogeneous of degree one, $\tilde{\pi}$ is degree-one homogeneous in $\mathbf{z}$. Let $\tilde{\pi}(\mathbf{z}, x)$ be the stochastic, before-tax profit if the plant is located at x and input vector $\mathbf{z}$ is used.

Let $(\hat{\mathbf{z}}, \hat{x})$ be optimal for (PL7.2) with tax rate $\hat{t}$, and notice that $((1 - \hat{t})\hat{\mathbf{z}}, \hat{x})$ is a feasible solution for any tax rate. Then

$$\begin{aligned} EU[(1 - \hat{t})\tilde{\pi}(\mathbf{z}^*/(1 - \hat{t}), x^*)] &= EU[\tilde{\pi}(\mathbf{z}^*, x^*)] \geqslant EU[\tilde{\pi}((1 - \hat{t})\hat{\mathbf{z}}, \hat{x})] \\ &= EU[(1 - \hat{t})\tilde{\pi}(\hat{\mathbf{z}}, \hat{x})] \end{aligned} \tag{7.25}$$

The equalities in (7.25) follow from the degree-one homogeneity of $\tilde{\pi}$ in $\mathbf{z}$, and the inequality follows from the optimality of $(\mathbf{z}^*, x^*)$ for $t = 0$. So $(\mathbf{z}^*/(1 - \hat{t}), x^*)$ must be optimal for (PL7.2) with tax rate $\hat{t}$. Furthermore, the weak inequality in (7.25) must hold as an equality, and so $U^*(0) = U^*(\hat{t})$. □

Theorem 7.3 holds for any number of inputs, not just for the case of two inputs, and any number of the input prices may be random.

The inconsistency of F being degree-one homogeneous and $\pi(P_1)$ being nonnegative for all P_1, can be explained in economic terms. If F is degree-one homogeneous, (7.24) states that the profit that occurs if price P_1 is realized can be written as the product of the marginal profit from input 1, $(P_0F_1 - P_1)$, and the amount of input 1 used. If $(P_0F_1 - P_1) \geqslant 0$ for all P_1, then increasing z_1 (with z_2 increased proportionately so z_1/z_2 and thus $F_1(z_1, z_2)$ are constant) would result in higher profit for those values of P_1 where $(P_0F_1 - P_1) > 0$ and no change for those P_1 where $(P_0F_1 - P_1) = 0$. Then, regardless of the firm's risk preferences (as long as $U' > 0$), the firm's *expected* utility would increase by increasing z_1 (and z_2). In effect, if F is degree-one homogeneous, a finite optimum will exist only if $\pi(P_1)$ is negative for some P_1 of positive measure.

Another way of viewing this result is to recognize that if F is degree-one homogeneous, *any* after-tax profit distribution achievable by the firm with no tax is also achievable for any tax rate $0 < t < 1$. Specifically, let $(\mathbf{z}^*, x^*)$ be the solution that produces the optimal profit distribution without taxes:

$$\tilde{\pi}^* = [\tilde{p}_0 - c_0(x^*)]F(\mathbf{z}^*) - \sum_{i=1}^{n} [\tilde{p}_i + c_i(x^*)]z_i^*$$

If an income tax of rate t is imposed, the firm's after-tax profit distribution can be maintained at $\tilde{\pi}^*$ by increasing all input levels to $z_i^*/(1 - t)$ and output to $z_0^*/(1 - t)$ and keeping the location fixed at x^*:

$$(1 - t)\left\{[\tilde{p}_0 - c_0(x^*)]F\left(\frac{\mathbf{z}^*}{1 - t}\right) - \sum_{i=1}^{n}\left[\frac{(\tilde{p}_i + c_i(x^*))z_i^*}{(1 - t)}\right]\right\} = \left[\frac{1 - t}{1 - t}\right]\tilde{\pi}^*$$
$$= \tilde{\pi}^*$$

So in this case, where delivered prices are not affected by the quantities z_0 and z_i, the imposition of a spatially uniform, flat-rate income tax does not affect the firm's after-tax profit distribution; it just causes more output to be produced to pay the taxes. It is crucial to recognize and appreciate this latter assumption (that all prices and transport rates are independent of the quantity of inputs and outputs used, produced, or shipped), because theorem 7.3 applies only over ranges of production where this assumption and the degree-one homogeneity assumption are valid. If these are ignored serious policy errors could result. For example, assuming that governments find it desirable to increase firms' output (and their use of labor) an apparent (but incorrect) conclusion is that the government should impose an income tax with a rate as close to one as possible; this would generate *expected* tax revenues approaching infinity, and employ all available labor without hurting the firm. However, the assumption of quantity-independent prices (perfectly competitive markets) requires that the amount of inputs used and output produced be small enough to not affect prices. If output is driven up very far this assumption would be violated. Furthermore, although F may be degree-one homogeneous over some range of production, as output level increases decreasing returns to scale usually occur.

Rather precise results concerning the effects of a flat-rate income tax can be obtained for general homogeneous production functions.

Corollary 7.2

(a) If $R'(\pi) > 0$ and F is homogeneous of degree $\lesseqgtr 1$, then z_1^*/z_2^* is (increasing,constant,decreasing) in t, (b) If $R'(\pi) < 0$ and F is homogeneous of degree $\lesseqgtr 1$, then z_1^*/z_2^* is (decreasing,constant,increasing) in t.

Proof: The case of degree-one homogeneity just repeats theorem 7.3; proof of the other two cases is similar to the proof of corollary 7.1, and is omitted. □

Comparison Between Lump-Sum and Income Taxes

Theorem 7.2 and corollary 7.2 provide contrasting results to those provided by theorem 7.1 and colollary 7.1. Under normal conditions (i.e., F exhibits decreasing returns to scale, and $U(\pi)$ has $r'(\pi) < 0$, and $R'(\pi) > 0$, a spatially uniform lump-sum tax imposed on the firm depicted in (PL7.2) leads to a reduction in input and output levels along with partial substitution away from the price-uncertain input and a tendency to shift the optimal plant location accordingly. A spatially uniform income tax imposed on the same firm leads to increases in input and output levels, an increase in the relative use of the price-uncertain input, and a locational adjustment opposite to that caused by a lump-sum tax.

In order to illustrate the main points contained in the two theorems and the two corollaries, consider the results for four cases for a risk-averse firm with declining absolute risk aversion, $r'(\pi) < 0$, increasing relative risk aversion, $R'(\pi) > 0$, and F is homogeneous of degree <1. The first case is that of certainty in all prices (with uncertain prices fixed at their expected values) with or without an income or a lump-sum tax. Optimal input and output levels in this case are denoted z_i^c $(i = 0, 1, 2)$. The second case incorporates uncertainty in the price of the first input with no taxes; optimal input and output levels are denoted z_i^u $(i = 0, 1, 2)$. The third case is that of uncertainty in the price of the first input with a lump-sum tax where the optimal values are z_i^T $(i = 0, 1, 2)$. The fourth case is that of uncertainty in the price of the first input with a flat-rate income tax; the optimal values are denoted z_i^t $(i = 0, 1, 2)$. To summarize:

Case	*Tax*	*Uncertainty*	*Variable*
1	with or without	none	z_i^c
2	none	$\tilde{p}_1$	z_i^u
3	lump-sum	$\tilde{p}_1$	z_i^T
4	income	$\tilde{p}_1$	z_i^t

Combining the results from theorems 7.1 and 7.2 and corollaries 7.1 and 7.2 we have

$$z_0^c > z_0^t > z_0^u > z_0^T \tag{7.26a}$$

$$\frac{z_1^c}{z_2^c} > \frac{z_1^t}{z_2^t} > \frac{z_1^u}{z_2^u} > \frac{z_1^T}{z_2^T} \tag{7.26b}$$

In considering (7.26a) and (7.26b) notice that the results for the uncertainty case with the flat-rate income tax are closer to the certainty results than are the results for the uncertainty case with no tax at all. In this respect, the flat-rate income tax may be considered to be an efficiency-inducing tax both technologically and spatially whereas a lump-sum tax is efficiency reducing. (This result may not be true for a progressive-rate income tax. Lippman and McCall [1982] have shown that under some conditions a progressive income tax can reduce investor risk taking and decrease efficiency. Whether or not the Lippman and McCall results apply to the models used here is still an open question.)

The imposition of a lump-sum tax shifts the distribution of $\tilde{\pi}$ to the left by the amount T. The spread or variation in $\tilde{\pi}$ is unchanged while the expected profit after taxes is lower and the firm considers itself to be poorer after the tax. This suggests that properties of the absolute risk-aversion function, $r(\pi)$, will be relevant. With $r'(\pi) < 0$ the firm has been moved by the tax to a more risk-averse portion of its utility function and it behaves as more risk-averse. Accordingly it uses less of the price-uncertain input both in relative and in absolute terms than it did prior to the imposition of the lump-sum tax. In contrast, when $r'(\pi) = 0$ the reduction in expected profit resulting from the imposition of the lump-sum tax does not influence the firm's risk preferences and thus has no impact on the firm's solution.

By way of contrast, the flat-rate income tax reduces both the size (or spread) of the variation in profit and the expected after-tax profit proportionally, suggesting that in this case the properties of the relative risk-aversion function, $R(\pi)$, are relevant. When relative risk aversion decreases as profit decreases (i.e., $R'(\pi) > 0$) the firm is more willing to accept risk after the imposition of the flat-rate income tax and consequently it increases its absolute and relative use of the price-uncertain input. The key factors are the assumptions that normally $r'(\pi) < 0$, $R'(\pi) > 0$ and the observation that the lump-sum tax impacts only the expected profit and not its "variability" while the flat-rate income tax affects the expected profit and its "variability" in proportion. These results with respect to the flat-rate income tax are consistent with the findings of other investigators, such as Domar and Musgrave [1944], Mossin [1968], Baron [1970], and Mayshar [1977].

As mentioned above, these results may seem to suggest that if governments find it desirable to impose policy measures that lead to an increase in firms' output (and their use of labor and other inputs), they should impose very high-rate income taxes. However, this is certainly not a proper conclusion. The assumption that input and output prices are exogeneous (though random) and independent of quantity used or produced (i.e., perfectly competitive markets) requires that the quantities of inputs used and output produced by a firm be small relative to the markets. If output rates and input usage are increased dramatically, this requirement will be violated. Other assumptions upon which our analysis rests will most likely also be violated in the face of extreme policy measures.

Other Taxes

Available analytical results for taxes other than lump-sum and flat-rate income are less complete and less decisive. Nevertheless, some meaningful results for an output tax of t^o per unit of output or an input tax of t^j per unit of input j are available and are presented below in the form of two theorems. The proofs of these theorems are similar in strategy to those of theorems 7.1 and 7.2 and are omitted here.

Theorem 7.4

In problem (PL7.2) with output tax t^o, let the location, x, be fixed. Then dz_1^*/dt^o, dz_2^*/dt^o, and $dz_0^*/dt^o < 0$ if $r'(\pi) \leqslant 0$, and ambiguous otherwise.

Theorem 7.5

In problem (PL7.2) with input tax t^j per unit on input j, let the location, x, be fixed. Then dz_1^*/dt^j, dz_2^*/dt^j, and $dz_0^*/dt^j < 0$ if $r'(\pi) \leqslant 0$, and ambiguous otherwise.

Theorems 7.4 and 7.5 state that under the usually assumed condition that $r'(\pi) \leqslant 0$, output taxes and input taxes both lead to reductions in input and output levels. Notice that theorem 7.4 also holds when the output tax is a tax on revenues generated from sale of output, that is a sales tax. (Refer to the earlier discussion and comparison of equations (7.1c) and (7.1d).) Theorems 7.4 and 7.5 say nothing about the impact of output, input, and sales taxes on the optimal input ratios. Indeed, results regarding the impact of t^o and t^2 on these ratios are ambiguous. The ratios are altered by these taxes if $z_1^* > 0$ and $z_2^* > 0$, but the direction of change in the ratio is not yet clear. Computational experience, using production relationships exhibiting constant or decreasing returns to scale, indicates that z_1^*/z_2^* increases with an input tax t^2, and that z_1^*/z_2^* is increasing in t^o if $R'(\pi) > 0$.

For an input tax on the price-uncertain input, t^1, more precise results can be obtained if the production function, F, is homogeneous. Using an analysis similar to that employed with the lump-sum tax and the flat-rate income tax it can be shown that

$$(dz_1^*/dt^1)/(dz_2^*/dt^1) = -F_{22}/F_{21} \tag{7.27}$$

This leads to corollary 7.3.

Corollary 7.3

If $r'(\pi) \leqslant 0$ and F is homogeneous of degree $\lesseqgtr 1$, then z_1^*/z_2^* is (decreasing, constant,increasing) in t^1.

Explanation of the Degree-One Homogeneous Case. The first-order conditions for (PL7.2) with an input tax, t^1, are:

$$E[U'(P_0F_1 - \tilde{P}_1 - t^1)] = 0 \tag{7.28a}$$

$$E[U'(P_0F_2 - P_2)] = 0 \tag{7.28b}$$

With $U' > 0$, Equation (7.28b) reduces to

$$P_0F_2 - P_2 = 0 \tag{7.28c}$$

As t^1 is perturbed with location, x, fixed, P_0 and P_2 do not change and consequently for (7.28c) and (7.28b) to hold, F_2 cannot change with t^1. When the production function, F, is degree-one homogeneous F_2 is constant along a ray from the origin. Of course, the input ratio, z_1/z_2, is also constant along a ray. Therefore, the input ratio, $\bar{z}_1/\bar{z}_2$ which is optimal for a given tax rate, $\bar{t}^1$, will be optimal for all levels of t^1. This is accomplished by altering the levels of z_1 and z_0 which, in turn, causes the effective price of input 1 ($\bar{P}_I + t^1$ + risk premium) to remain unchanged and so the optimal input ratio remains constant. Specifically, if t^1 is increased by Δt^1, z_1^* will be decreased (and z_2^* proportionately) just enough so that the risk premium decreases by Δt^1, which keeps the effective price and input ratio constant.

The result for t^1 reported in corollary 7.3 when the production function exhibits decreasing returns to scale is expected. However, with constant returns to scale, the input tax is neutral with respect to the optimal input ratio (mix). This result, as can be deduced from the discussion in the preceding paragraph, depends on P_2 being deterministic. (If P_2 is not deterministic, the result does not hold.) This latter result is quite surprising. We have made the point that many taxes that are neutral with respect to input mix and location are not neutral when price uncertainty is considered, except under very special conditions. In contrast, the input tax, which is never neutral under certainty, can be neutral under uncertainty, but only under very special conditions.

Policy Implications

Advantages of Flat-Rate Income Tax

We have demonstrated that, with risk-averse decision makers and input price uncertainty, the imposition of a flat-rate income tax can be expected to increase the firm's willingness to undertake risk. Further, it was shown—see (7.26a) and (7.26b)—that an income tax will usually push the firm's optimal solution closer to its deterministic solution (i.e., the optimal solution with all random variables set at their expected values). We investigate the policy implications of these results below.

Consider a firm located at site x and facing problem (PL7.2). We will assume that $r'(\pi) < 0$ and $R'(\pi) > 0$, although formally we require strict inequality only for one of the two measures of risk aversion. As discussed in chapter 5, these assumptions correspond to what are usually considered the normal cases. Denote the optimal production decision with p_1 random and no taxes imposed as (z_0^u, z_1^u, z_2^u). We will write: $\tilde{p}_1 = \bar{p}_1 - \tilde{\varrho}$, where $E(\tilde{\varrho}) = 0$. Then the optimal profit distribution can be written as

$$\tilde{\pi}^u = [p_0 - c_0(x)]F(z_1^u, z_2^u) - [\bar{p}_1 - \tilde{\varrho} + c_1(x)]z_1^u - [p_2 + c_2(x)]z_2^u \tag{7.29a}$$

$$= \bar{\pi}^u + \tilde{\varrho} z_1^u \tag{7.29b}$$

where $\bar{\pi}^u$ is the expected profit.

In a similar way let (z_0^T, z_1^T, z_2^T) be the optimal solution when a lump-sum tax of T is imposed. The resulting optimal before tax profit, $\tilde{\pi}^T$, can then be written as

$$\tilde{\pi}^T = \bar{\pi}^T + \tilde{\varrho} z_1^T \tag{7.30}$$

Finally, let (z_0^t, z_1^t, z_2^t) be the optimal solution when a flat-rate income tax of t is imposed such that $t = T/\bar{\pi}^T$. The resulting before-tax profit $\tilde{\pi}^t$ is

$$\tilde{\pi}^t = \bar{\pi}^t + \tilde{\varrho} z_1^t \tag{7.31}$$

(Recall that x is fixed at the same location for all three cases.) From (7.26a) we know that $z_0^t > z_0^u > z_0^T$. Now from the optimality of (z_0^t, z_1^t, z_2^t) under income tax t, we know

$$EU[(1 - t)\tilde{\pi}^t] > EU[(1 - t)\tilde{\pi}^T] \tag{7.32a}$$

or equivalently,

$$EU[(1 - t)\bar{\pi}^t + (1 - t)\tilde{\varrho}z_1^t] > EU[(1 - t)\bar{\pi}^T + (1 - t)\tilde{\varrho}z_1^T] \tag{7.32b}$$

From theorems 7.1–7.3 and their corollaries we deduce that $z_1^t > z_1^T$ and therefore the random variable $(1 - t)\tilde{\varrho}z_1^t$ has a uniformly larger spread than the random variable $(1 - t)\tilde{\varrho}z_1^T$. From the results of Rothschild and Stiglitz [1970], the inequality (7.32b) can hold only if $\bar{\pi}^t > \bar{\pi}^T$. (That is, if some gamble $\tilde{A}$ has a uniformly greater risk (i.e., larger spread) than another gamble $\tilde{B}$, then a risk-averse decision maker will prefer $\tilde{A}$ to $\tilde{B}$ only if the expected value of $\tilde{A}$ is greater than the expected value of $\tilde{B}$.) Now consider the term on the right-hand side of the inequality (7.32b),

$$\begin{aligned} EU[(1 - t)\bar{\pi}^T + (1 - t)\tilde{\varrho}z_1^T] &= EU[\bar{\pi}^T - T + (1 - t)\tilde{\varrho}z_1^T] \\ &> EU[\bar{\pi}^T - T + \tilde{\varrho}z_1^T] \\ &= EU[\tilde{\pi}^T - T] \end{aligned} \tag{7.33}$$

The inequality in (7.33) is because $\bar{\pi}^T - T + (1 - t)\tilde{\varrho}z_1^T$ has a uniformly smaller spread than $\bar{\pi}^T - T + \tilde{\varrho}z_1^T$ but has the same expected value.

From (7.32b) and (7.33) we conclude that the firm would prefer the solution associated with the flat-rate income tax to the solution associated with the lump-sum tax. This is true even though the expected value of taxes paid under the income tax, $t\bar{\pi}^t$, is greater than with the lump-sum tax $t\bar{\pi}^T = T$. If the government collecting the taxes were risk-neutral, as is often assumed, it too would prefer the uncertain revenue, $t\tilde{\pi}^t$, from the income tax to the certain amount, T, from the lump-sum tax because the expected revenue is larger. Even if the government were risk-averse it might still prefer the uncertain revenue from the income tax. Because a government is likely to have a number of sources of income, the variability in any one source is generally less important and there is the possibility that risks from the business income tax may be negatively correlated with risks associated with other governmental income sources. Similar arguments can be utilized to contrast the income tax with the output and input taxes.

If governments can become aware of the risks faced by firms within their jurisdiction and of their risk preferences, they can use this information to promote economic activity through the tax structure while at the same time increasing their expected revenue. Governments employing such strategies ought to have a competitive advantage in attracting and keeping desirable firms over those governments and regions that do not consider risk preferences in formulating their tax structures.

Numerical Example

A numerical example will now be used to illustrate the preceding results.

Example 7.1

Consider a firm that produces a single output using two inputs via a Cobb-Douglas (homogeneous) production function that exhibits decreasing returns to scale: $z_0 = F(z_1, z_2) = z_1^{0.48} z_2^{0.48}$. The sources of the inputs are at distinct locations one distance-unit apart. For convenience, let city A be the source of input 1 and city B be the source of input 2. The price of the first input at its source is $\tilde{p}_1 = 3.5 - \tilde{\varrho}$. The distribution of $\tilde{\varrho}$ is uniform with $\text{Prob}(\varrho = k) = 0.2$ for $k = -2, -1, 0, 1, 2$. The price of the second input at its source is $p_2 = 4.1$ and the price of output is $p_0 = 9.5$. The transport cost is 0.5 per unit of input per unit of distance and the transport cost of output is ignored. Let the firm's utility function be $U(\pi) = -\exp(-0.25\pi)$. In this case, $r(\pi) = 0.25$, $R(\pi) = 0.25\pi$, $r'(\pi) = 0$, and $R'(\pi) = 0.25 > 0$. The analysis developed in chapter 5 shows that under the conditions of this example the optimal location will be at an endpoint of the location line—that is at city A or at city B.

If there were no price uncertainty and $\tilde{p}_1 = \bar{p}_1 = 3.5$, the optimal solution for the firm is to locate at city A and produce output at the rate $z_0^c = 7.120$ and use inputs at the rates $z_1^c = 7.984$ and $z_2^c = 7.478$. However, under uncertainty with no taxes the optimal location is at city B. Under these circumstances the optimal rate of production is $z_0^u = 1.86$, and optimal input utilization rates are $z_1^u = 1.762$ and $z_2^u = 2.068$ with expected utility of -0.7049.

Suppose that the governments of cities A and B wish to raise at least 0.50 in taxes from the firm should it located within their jurisdictions. If each city were to impose a lump-sum tax there would be no change from the no-tax solution (z_0^u, z_1^u, z_2^u, B) since $r'(\pi) = 0$. Instead, suppose that city A were to impose an income tax of rate $t = 50$ percent while city B retains the lump-sum tax. Then EU at A is -0.7865 while EU at $B = -0.7988$, so the firm's optimal location (refer to table 7–1) is now at city A with $z_0^* = 2.60$, $z_1^* = 2.840$, $z_2^* = 2.578$. By selecting its tax carefully, city A is not only able to attract the firm to its region, but the firm's

Table 7–1. Optimal Solutions for Example 7.1

$$U(\pi) = -e^{-0.25\pi}$$

	Optimal Production Values at A					Optimal Production Values at B				
	z_0	z_1	z_2	*Expected Utility*	*Tax Paid*	z_0	z_1	z_2	*Expected Utility*	*Tax Paid*
No tax	1.70	1.790	1.688	−0.7129	0	1.86	1.762	2.068	−0.7049	0
Lump-sum tax = 0.5	1.70	1.790	1.688	−0.8078	0.5	1.86	1.762	2.068	−0.7988	0.5
Income tax 50%	2.60	2.840	2.578	−0.7865	1.45	2.80	2.740	3.118	−0.7845	1.43

$$U(\pi) = -e^{-0.15\pi}$$

	Optimal Production Values at A					Optimal Production Values at B				
	z_0	z_1	z_2	*Expected Utility*	*Tax Paid*	z_0	z_1	z_2	*Expected Utility*	*Tax Paid*
No tax	2.32	2.514	2.297	−0.7684	0	2.52	2.444	2.806	−0.7649	0
Lump-sum tax = 0.5	2.32	2.514	2.297	−0.8282	0.5	2.52	2.444	2.806	−0.8245	0.5
Income tax 50%	3.52	3.944	3.489	−0.8327	1.79	3.74	3.758	4.155	−0.8338	1.73

output level is higher than with a lump-sum tax (or without *any* tax) as is its use of the input supplied at A, input 1, and this may produce additional economic benefits to city A. Furthermore, the expected value of tax revenue from the income tax is $0.50(\bar{\pi}') = 1.45$, which is almost three times the amount that would have been raised with the lump-sum tax. City A could use a flat-rate income tax rate anywhere in the range of 20 percent to 55 percent and still make city A more desirable than city B with its 0.50 lump-sum tax, while smultaneously producing expected tax revenue greater than 0.50.

The taxing authority at city B could counter the actions of city A by imposing an identical flat-rate income tax in lieu of the lump-sum tax. Under these circumstances, the firm's optimal location (refer again to table 7–1) is city B with $z_0^* = 2.80$, $z_1^* = 2.740$, $z_2^* = 3.118$. However, it is not always the case that a location that is optimal under uncertainty with no taxes is optimal under uncertainty when both competing cities employ identical income taxes. The results presented in the preceding sections suggest that if the same tax (in rate and form) is imposed everywhere, the firm's optimal input mix, output level, and thus its location, may still depend on the form of the tax. Again refering to table 7–1, suppose for example that the firm's utility function is $U(\pi) = -\exp(-0.15\pi)$. Then with either no tax or a lump-sum tax of 0.50 at both cities, the firm's optimal location is at B with $z_0^* = 2.52$, $z_1^* = 2.444$, $z_2^* = 2.806$. However, if both cities were to employ a flat-rate income tax of 50 percent, the firm's optimal location is city A with $z_0^* = 3.52$, $z_1^* = 3.944$, and $z_2^* = 3.489$.

Within the scenarios described by problem (PL7.2), and under the assumption that $U'(\pi) > 0$, $U''(\pi) < 0$, $r'(\pi) \leqslant 0$, $R'(\pi) \geqslant 0$ and decreasing returns to scale, for any lump-sum tax there is always an income that is more desirable to the firm yet produces greater expected tax revenue. Depending upon the government's portfolio of risks and its own risk preferences, it is likely that the income tax would be preferred by the government as well. However, the income tax may not be equally desirable to all governments because of differences in risk preferences and in risk portfolios, as well as differences in secondary benefits or costs associated with the tax. For example, governments that are less risk-averse or have a better portfolio of income and risks will be better able and more likely to exploit income taxes to attract firms. Further, if the relative risk-aversion function has positive slope (i.e., $R'(\pi) > 0$) and if the production function, F, exhibits decreasing returns to scale, the imposition of the income tax will lead to substitution toward the price-uncertain input. Secondary benefits then accrue to those cities that are potential locations and close to the source of the price uncertain input.

Concluson

The material presented in this chapter has shown that when factor prices are random variables the imposition of taxes, even if spatially uniform, can have significant effects on the optimal input mix, location, and output rate of a firm attempting to maximize a von Neumann-Morgenstern utility function of profit. Spatially uniform taxes, which are technologically and spatially neutral under certainty, were shown to be nonneutral when factor prices are random variables, and firms are risk-averse. This contradicts the conventional wisdom; for example, Stafford [1980, p. 110] writes, "Regardless of their magnitude, taxes which do not exhibit spatial variation are unimportant in deciding between alternative locations, e.g., if all prospective sites are within the U.S., then Federal taxes are not locationally significant.... Always, the key locational issue is the differential from place to place." The results developed in this chapter demonstrate that spatially uniform taxes *do* matter in choosing among alternative sites; spatially uniform taxes can have spatial biases which favor one site over another. (In fact, these biases can counteract the effects of spatially differential taxes, thereby, leading to apparently anomolous location behavior by firms.)

In addition, it was shown that the form of tax imposed may have as important an effect on the firm's production and location decision as the expected amount of tax revenue paid. Thus two forms of taxation that raise the same revenue may not be equally desirable to either the firm or the government. Specifically, because a flat-rate income tax causes the government to share the risk of input-price uncertainty with the firm, it can be designed to generate greater expected tax revenue for the government than other common forms of taxes and yet provide a more desirable tax environment for firms.

In addition to the obvious policy implications, this result has considerable relevance for empirical research on the spatial effects of tax differentials. Just as Fox [1981] found that zoning restrictions confound the spatial effects of tax differentials, our results suggest that the mix of tax forms may mask the effects of the total tax levy. Thus, the form and mix of taxes should be considered in any empirical research. Furthermore, tax action at the national level may have unexpected regional biases, not only because the mix of firms differs among regions, but under price uncertainty spatially uniform taxes can alter the effective prices of inputs in a spatially differential manner.

State and local governments have drawn criticism for their competitive use of tax policy to attract and to promote the growth of firms. The major criticisms seem to fall into three primary categories: (1) such policies hurt overall economic efficiency by distorting the optimal locations, production levels, and input mixes; (2) if all governments use these policies the governments as a whole hurt themselves by losing tax revenues; and (3) local governments are often exploited by firms that extract tax concessions, yet would have located within the local government's jurisdiction anyway. The criticisms are all directed toward tax concession policies. We have shown that these criticisms may not apply to the *imposition* of a flat-rate income tax; in fact, the opposite may be true. Income taxes imposed when there is price uncertainty may induce firms to select output rates, input mixes, and facility locations closer to the deterministic optima (i.e., closer to the solution of the deterministic problem in which the random price is set at its expected value), than do other taxes. Furthermore, governments can make their regions more attractive by changing the form or mix of taxes rather than by lowering rates. As we have shown, if well designed, income taxes can generate greater expected tax revenue for the governments as a whole, and eliminate the opportunity for exploitation by firms.

In this chapter the production/location framework was used to illuminate many technological and spatial consequences of tax alternatives. While useful for providing insights into the effects of tax policy, a model exclusively at the firm level has little to say regarding questions of tax incidence, interindustry effects, labor migration, and collection efficiency. The answers to these questions must await the extension of the partial equilibrium model to a regional and general spatial equilibrium model.

Notes

1. Theorem 7.1 is stated in a strong form which assumes $z_1^* > 0$ and $z_2^* > 0$. If either z_1^* or z_2^* equals zero, the inequalities in part (a) are weak.

2. Note that under uncertainty a finite optimum can exist even if the production function exhibits constant or increasing returns to scale.

3. In the original version of this theorem in Martinich and Hurter [1985], the condition $\pi(P_1) \geqslant 0$ for all P_1 was omitted. Martinich and Hurter [1987] corrects this.

4. $R(P_1) = -\pi(P_1)U''(\pi(P_1))/U'(\pi(P_1))$ and $U''(P_1)/U'(P_1) > 0$ for all P_1 (assuming that $U'' < 0$ so as to insure a finite optimum). Then $0 < P_1 < P_1^0$ implies that $\pi(P_1) > 0$, so $R(P_1) > 0$.

8 POLICY EVALUATION: INCENTIVES AND REGULATION

Much has been written about the importance of incentives and regulations in the site selection decision making of firms. However, very few papers and books on these subjects treat the potential impact of parametric (environmental) uncertainty on such decisions. Furthermore, we are able to find only a single published work that uses the production/location framework to evaluate government regulation, and only our own work uses this framework to evaluate locational incentives. This is unfortunate in view of James B. Cannon's admonition that, "choice of location cannot be identified as an *independent* characteristic of a strategic investment decision.... The impact of locational incentives must be seen in the broader context of the total investment decision" [Collins and Walker 1975, p. 112].

Conceptually, incentives and regulations could be differentiated according to their purpose. On the one hand there are, for example, regulations, such as environmental regulations, the goal of which is to reduce pollution but not to influence locational choices. On the other hand, there are regulations and incentives directly intended to influence the locational choices of firms. However, such a differentiation is not necessarily valid; spatially uniform taxes are generally not intended (and often not believed) to affect locational choices. But, as demonstrated in chapter 7, they can have a significant locational impact nevertheless. Although very little formal analysis using a production/location framework has been done for incentives and regulations, the work available indicates that the former type of incentive/regulation (i.e., not intended to affect location choices) *can* have significant spatial effects due to the interactions among the production and location variables.

We first discuss the spatial effects of incentives. We begin with a brief review of recent empirical work and put those findings in perspective and then discuss tax abatements and subsidies. There has been no formal published work on these incentives using the production/location framework, but because of their similarity to lump-sum taxes, we are able to propose conjectures concerning their likely spatial effects. To conclude the first section of this chapter, we propose and illustrate an interesting form of locational incentive based upon explicit risk-sharing/reduction. The most promising and unique feature of this incentive scheme is that the governmental entity offering the incentive can provide a very attractive incentive, while simultaneously generating revenue rather than giving up revenue, as with traditional incentive schemes. In the second section, regulations are discussed. Environmental and rate-of-return regulations are featured because they are such common forms of regulation, and the latter has been the primary focus of regulatory research.

Incentives

The extent to which locational incentives (and tax differentials) affect the locational choices of firms has been an issue for debate for the past 30 years. For the most part empirical studies have been rather mixed in their findings concerning the effects that tax abatements and other financial incentives have on firms' locational decisions. Observational studies of firms' locational behavior have generally found little or no effect, while questionaire studies have indicated that such incentives are definitely a consideration in firms' analyses. Schmenner [1982, p. 51] summarizes his empirical findings as follows: "[T]he foregoing results demonstrate fairly convincingly that tax and financial incentives have little influence on almost all plant location decisions. Taxes and financial inducements seem to be, at best, tie-breakers acting between otherwise equal towns and sites." And yet, Schmenner, based upon his 1980 survey of *Fortune* 500 companies, reports that 71 percent of the firms responding received some form of governmental assistance. Physical help appears to be used more extensively than financial help. The *Fortune* 500 survey showed that 61 percent of the firms responding used at least one physical aid whereas only 31 percent used financial aids. The financial aids listed in the *Fortune* 500 survey were (a) industrial revenue bonds, (b) industrial revenue bonds for pollution control, (c) tax concessions of any sort, and (d) loan guarantees. The physical aids included: (a) roads, sewers, and water, (b) labor training,

(c) help with environmental permits, (d) zoning changes, (e) expansion of sewage treatment facilities, and (f) traffic and parking adjustments.

The empirical studies that claim financial inducements and tax abatements have no substantial effect on locational decisions appear to be at odds with this high reported usage of governmental "aids". It is even more at odds with the well-publicized competition among states in the U.S. to attract industrial facilities such as auto and computer plants, and research centers. This apparent inconsistency may be due to at least three factors.

1. In a large majority of cases, the economic differences between competing locations is so large that government inducements could not reasonably be expected to make an inferior location preferred by the firm. As Schmenner states, it is only in the marginal cases, where two locations are relatively equivalent (which may be 5–10 percent of all locational decisions), where differentials in government inducements make a difference. Thus, the empirical studies that try to correlate the observed location choices (revealed preferences) for *all* firms versus spatial differentials in inducements are unlikely to capture the effects on these marginal decisions because they are lost in the sea of nonmarginal cases. This explains why empirical studies of the preceding type rarely have found incentives to have significant spatial effects, while surveys of the type Schmenner performed report many firms using government aids.

2. Many empirical studies have focused upon the effects of only one or two types of incentives, and they tend to be financial incentives, such as tax abatements or subsidies. Yet these are only a small part of the overall package that governments offer. In fact, Schmenner's survey indicates that physical rather than financial inducements are used more extensively. These physical inducements are just as real and may be more attractive and valuable to firms. Thus, Schmenner [1982, p. 58] recommends, "States and localities should stand ready to offer the interested manufacturers (i) speedy and accurate information about potential sites, (ii) help in securing necesssary environmental and zoning permits, (iii) timely help with roads, sewage, water, waste treatment, and labor training which can make a real difference to the smooth startup of a new manufacturing facility."

3. Many empirical studies have looked only at differentials in total taxes paid or financial inducements received and have not considered the form of inducement. As shown in chapter 7, taxes that generate the same revenue can have different attractiveness for firms and the same tax can have differential spatial effects, especially within a stochastic framework. The same holds true for many incentives. Empirical studies have tended to ignore both the effects of form and the role of uncertainty and risk preferences in locational studies.

Our conclusion is that the shortage of empirical evidence supporting locational incentives is due to methodological and informational deficiencies, rather than a true absence of effect. In fact, the production/location framework makes it possible to identify the potential spatial effects of government policies which are not intended to be overtly spatial in their effect, as well as those intended to be locational inducements. Results from such theoretical models and analysis should provide better guidance for future empirical research; indeed, it may be the absence of such a theoretical framework that has caused the deficiencies in earlier empirical work in this area. It is for this reason that we have included the following preliminary work in this volume.

Subsidies and Tax Abatements

To a large extent most forms of subsidies used by governments to attract industry can be treated as being very similar to negative taxes. In some instances, the subsidies are fixed amounts (e.g., cash payment, or more commonly, free land, free roads, free sewer connection) comparable to negative lump-sum taxes. In other cases the subsidies may take the form of negative input taxes, such as providing free training for employees or reduced electricity rates. As such, the production and spatial effects of subsidies and tax abatements are generally opposite those derived for taxes in chapter 7. Consequently, we will devote little space to this topic since the reader can deduce all the relevant results directly from chapter 7. However, we do want to emphasize the point that under uncertainty and risk aversion, spatially uniform subsidies (e.g., at a national level) can have a technologically and spatially nonneutral effect on firms; that is, they may favor one region over another. To illustrate this point we present the following example.

Example 8.1

Consider problem (PL7.2) with the objective function altered to reflect a lump-sum subsidy available at all locations (i.e., equation (7.7) has a subsidy of amount A added to it). Suppose the firm's production function is $F(z_1, z_2) = z_1^{0.48} z_2^{0.48}$. The price of the first input at its source location, $(0, 0)$, is $\tilde{p}_1 = 3.6 - \tilde{\varrho}$ where $\text{Prob}(\varrho = k) = 0.2$ for $k = -2, -1, 0, 1, 2$; the price of the second input at its source location, $(2, 0)$, is $p_2 = 4.0$; the price of the output at its market location, $(1, 1.732)$, is $p_0 = 10.0$. Let the per unit transport cost for all commodities be 0.25 per unit distance, and let the set of feasible locations be $S = R^2$ (actually the optimal location will lie within the location triangle formed by the source and market locations). Finally, let the firm's utility function be $U(\pi) = \pi^{0.5}$ (note that $r'(\pi) < 0$ and $R'(\pi) = 0$).

The firm's optimal solutions for various lump-sum subsidy levels are given in table 8–1a. As the subsidy level, A, increases, the firm's level of risk-aversion decreases (because $r' < 0$); its optimal output level, z_0^*, and input ratio, z_1^*/z_2^*, increase; and the optimal location moves toward the source of the first (random) input. The effect is even more dramatic if we treat the location space as a network connecting the three source and market points. From chapter 6 it follows that the optimal location will be one of the nodes, so that S can really be reduced to $S = \{(0,0), (2,0), (1, 1.732)\}$. The optimal solutions for various values of A are given in table 8–1b. Again notice that even though the same subsidy is offered at all locations, the magnitude of the subsidy can affect the optimal output level, input mix, and plant location. This is the type of phenomenon that previous empirical studies have not considered, and that could mask the effects of spatially differential subsidies and taxes.

Table 8–1a. Optimal Solutions When Feasible Space $S = R^2$

Subsidy A	*Optimal Location*	z_0^*	z_1^*	z_2^*	$EU(\tilde{\pi})$
5	(1.04, 0.53)	3.73	3.82	4.06	2.8074
10	(1.02, 0.53)	5.43	5.75	5.91	3.7059
20	(0.99, 0.52)	7.73	8.43	8.40	4.9723
30	(0.98, 0.51)	9.32	10.34	10.12	5.9448

Table 8–1b. Optimal Solutions When Feasible Space S = Market Connecting Network

Subsidy A	*Optimal Location*	z_0^*	z_1^*	z_2^*	$EU(\tilde{\pi})$
5	(2.0, 0.0)	3.48	3.39	3.96	2.7278
10	(2.0, 0.0)	4.99	5.00	5.69	3.6197
20	(0.0, 0.0)	6.55	7.56	6.63	4.8873
30	(0.0, 0.0)	7.90	9.27	8.00	5.8647

Risk-reducing Incentives

In chapter 7 we demonstrated that within a stochastic environment, income taxes are a preferred form of business taxation for many risk-averse firms because it makes the government a partner in the risk, thereby, reducing the risk faced by the firm. Risk-averse firms are often willing to pay, directly or indirectly, for a reduction in the risk they face. (This is the meaning of the so-called "risk-premium" introduced in chapter 5.) If risk reduction is viewed as beneficial by firms, even in the form of taxation, this suggests that more direct forms of risk reduction might be developed in place of subsidies and other locational incentives. Therefore, a relatively risk-neutral government (or third party) could absorb some or all of the price risk facing a firm using a price-uncertain input, receive a fee for doing so, and both parties would benefit. One mechanism for accomplishing this end is for a government to provide price insurance. For example, the government might act as an intermediary to purchase some of the price-uncertain input and subsequently to resell it to the firm at a guaranteed price equal to $p_1^* = \bar{p}_1 + \alpha$ where $\bar{p}_1$ is the anticipated average price for the input and $\alpha > 0$ is the premium paid to the government.

To illustrate this possibility we will again make use of example 7.1 We begin by assuming no taxes and that $U(\pi) = -e^{-0.25\pi}$. The firm's optimal solution is to locate at city B with production decisions $z_0^* = 1.86$, $z_1^* = 1.762$, $z_2^* = 2.068$. If the government at A wanted to attract this firm to its locale, the traditional approach would be to provide some form of subsidy such as a tax abatement, free land, or a less than market interest rate loan. In the example at hand, the government would have to give the firm a direct subsidy of approximately 0.05 to make location at city A more desirable than at city B.

As an alternative, the government at A could instead offer to act as a price insurer for the price-uncertain input (input 1). If the government at city A gauarantees input 1 at a price of 4.05 per unit (notice that the expected price is only 3.50), the firm's optimal location then shifts to city A with $z_0^* = 3.74$, $z_1^* = 4.208$, $z_2^* = 3.710$. (Of course, the firm would have to make a commitment to buy input 1 from the government even if the market price is less than 4.05 per unit; otherwise, the government would always lose.) Consequently, the government at city A would: (1) attract the firm to its locale, thereby increasing local employment and generating secondary benefits; (2) increase the firm's use of input 1 for which city A is the source, further increasing local economic benefits; and (3) receive an expected revenue of 0.55 per unit of input 1 used rather than having to pay the firm a subsidy.

Price insurance and risk reduction are especially attractive as locational incentives because they avoid many of the criticisms levied against more commonly used incentive structures. One frequent criticism is that many of the firms that take advantage of subsidy-based incentives would have located in the region offering the subsidy even if the subsidy were not present, thereby wasting the government's subsidy without any real gain to the government. With risk-reducing incentives, even if a firm accepts the incentives when it would have located within the risk absorbing government's locale without the incentive, the government loses nothing. If well designed, the government will generate net positive revenue from providing the insurance. Thus, a risk-reducing incentive acts as an attraction for firms that would benefit from such action, while firms that would locate within the government's jurisdiction anyway do not penalize the government. This means that the governmet does not need to determine a priori whether or not the target firm would locate within its jurisdiction before making such an offer.

Related criticism is that incentives often help the firms least in need of them and that it is difficult to discriminate among firms. However, to a large extent risk-reducing incentives have a bias toward smaller firms and they allow firms to "self-discriminate." Smaller, more risk-averse firms will usually find these risk-reducing incentives well worth the premium to them, whereas larger and more risk-neutral firms are less susceptible to risk-reducing incentives because the risk-absorbing feature is not worth the premiums to them. (In general, smaller more entrepreneurial firms usually have less risk-averse utility functions than larger more managerial firms over the entire range of profit levels. However, the smaller firms are working from a smaller base profit and wealth. Since the absolute risk-aversion function is decreasing for most firms ($r'(\pi) < 0$), the absolute risk aversion of small firms within their (lower) relevant profit range is often greater than that of large firms within their higher relevant profit range. Thus, risk-reducing incentives possess a natural discrimination favoring those firms that are more vulnerable to locational manipulation.)

A final related criticism is that locational incentives do not generate sufficient gain to justify their expense. But, if appropriately designed, risk-reducing incentives will not lose money; instead, they generate a profit for the government (over the long term) equal to the difference between the guaranteed price and the expected price times the volume of product "insured".[1]

Advantages of Risk-reducing Incentives

The main point of the preceding subsection is that if a government were to act as an intermediary for a risk-averse firm by buying price-uncertain inputs at the going (uncertain) price and reselling to the firm at a guaranteed price above the long-term average, the government could earn a profit over the long term, while the firm might find such an arrangement sufficiently attractive to locate within the government's jurisdiction. The example used to illustrate this point is certainly quite specific, so some qualifications and generalizations are in order.

First of all, even though a government could earn an expected profit by using a price-insurance scheme, on occasion the government might actually lose money in the short term due to price fluctuations (of course there are other economic benefits which might outweigh this loss). Consequently, it is not obvious that the government would prefer their incentive gamble to the alternative of a guaranteed "profit" of zero or even a guaranteed loss using classical incentives. If the government were less risk-averse than the firms it was targeting, then the government should find the risk-reducing incentive strategy attractive. But it is not unreasonable to believe that some governments (i.e., politicians) are at least as risk-averse as many firms. In this case these incentives may not be attractive. It is also possible that there may be other political or legal reasons why a government could not take such direct risk-reducing actions. In cases where direct action may be impossible, indirect actions might be available. For example, the government could encourage and support third-party efforts to reduce business risk. This might be done via creation of futures markets for some inputs or outputs, or a nongovernmental organization (e.g., Chamber of Commerce, AFL-CIO, or some regional business development organization) may be willing to absorb price risks. The risk absorption need not be complete in order to be effective. Actions such as the income tax, which share risk, may be very effective while limiting the government's risk. For example, in example 8.1 the government could sell input 1 for $3.5 - 0.5\tilde{\varrho} + \alpha$. The government's risk is also likely to be limited because these incentives would constitute only a small portion of a government's total revenue. The government may also want to promote and participate in other risk-reducing actions, such as long-term contracts for labor, utilities, or fixed tax rates. Indeed, guarantees of availability and *timing* for such essentials as roads, permits, sewers, may be considered risk-reducing incentives to firms. The ultimate costs to the firm (and government) may not be less than would be the case without such guarantees, but the removal or reduction of the risk is a major inducement.

The preceding discussion has focused upon using reduction in risk as a way to improve the business environment for businesses and thereby attracting economic activity. The opposite is, of course, also true. Government actions that increase risk tend to add a risk premium onto firms' cost of doing business. For example, when governments discuss the possibility of raising tax rates, pollution penalties, etc., not only do the expected costs from these actions increase, but there is an added risk premium to be considered by firms in their decision making until the issues are clearly resolved. Thus, the stochastic production/location models in this book can be used to demonstrate the old adage that known bad news may be less damaging to firms than uncertainty.

Regulations

Government regulations can take many forms. The primary purpose of a regulation may not have anything to do with firm location decisions; nevertheless, because of the inherent interrelationships between plant design and location variables, the regulation, acting through the plant design variables, can have location effects. For example, regulations controlling interstate commerce through transportation rates may affect not only the cost of commodities to the consumer but the location of new plants as well. Similarly, regulations that directly affect the production variables can have significant spatial effects. For example, regulations that restrict the use of some inputs (e.g., natural gas, water, labor) or the production of some outputs (e.g., dioxin) could affect plant location decisions, even when such regulations are applied at all locations. In the next subsection we will suggest ways in which input and output (environmental) regulations can be incorporated into production/location models. In the following subsection, rate of return regulations will be discussed. Although this form of regulation is very common and widely studied, the potential spatial consequences have received little attention.

Input and Environmental Regulations

Regulations that restrict the use of inputs usually can be incorporated into production/location models by adding constraints of the form $z_i \leq L_i$ where L_i is some allowable limit. Even if the limit is imposed everywhere, if the limit is binding then the firm is forced to alter its input mix and output level. This clearly changes the net locational pull and can cause the firm to alter its optimal location. The following variation of example 2.5 illustrates this point.

Example 8.2

Consider problem (DL2.1)—that is, the firm produces one output from two inputs; the source of input 1 is at 0, the source of input 2 and the output market are at point s, output transport costs are included, the output level is fixed, and the firm is minimizing cost. Let $p_1 = p_2 = 2$, $s = 2$, $c_0(x) = 1.5(2 - x)$, $c_1(x) = x$, $c_2(x) = (2 - x)$, $\bar{z}_0 = 12$ and $F(z_1, z_2) = z_1^{0.6} z_2^{0.2}$. The optimal solution is $x^* = 0$, $z_1^* = 34.955$, $z_2^* = 5.826$. If an imposed regulation restricts the use of input 1 to a maximum of 25.0 units, then the optimal solution becomes $x^* = 2$, $z_1^* = 24.720$, $z_2^* = 16.471$.

There are two important points to notice about these results: (1) even though the input restriction is imposed at all locations it causes the optimal input mix and location to change; and (2) at the new optimum location the input restriction is not binding! This latter result is very surprising and could cause an outside observer to conclude that the input regulation is innocuous. In fact, the opposite is true. The firm would like to locate at $x = 0$, and use large amounts of input 1 (34.955 units), but the input regulation forces the firm to use an input mix of $z_1 = 25.000$, $z_2 = 15.926$ if it locates there. This distortion of the input mix makes location $x = 0$ inferior to $x = 2$, where the optimal input mix, in this case, is not affected by the regulation. Thus, the production/location framework identifies another spatial anomaly. Unless the possibility of such anomalies is recognized, empirical research that simply looks at firms' final production and location choices can produce misleading results and conclusions.

Environmental restrictions can be incorporated into the production/location framework in at least three ways. First, in some cases the production of a pollutant can be tied directly to the usage of one or more inputs. A limit on the output of the pollutant would then be included in the models by adding a constraint of the form $g(z_i, \ldots, z_j) \leqslant L_i$ where g is the function that computes the amount of pollutant produced as a function of the relevant inputs. (Notice that the simple input constraints $z_i \leqslant L_i$ are a special case of this.) Second, there could be an explicit limit on output level: $z_0 \leqslant L_0$. This presents no analytical problem because, if the constraint is binding, it simply reduces the firm's problem to a design/location problem, which has been discussed extensively already.

The final possibility requires a different approach. The substances produced by a manufacturing plant may be so toxic that governments may explicitly prohibit the plant from being located within a certain distance of populated areas. In this case the set of feasible locations must be redefined to exclude prohibited areas. Conceptually, such a modification is straightforward, but from a practical viewpoint this can create many problems. For example, if initially the feasible set $S = R^2$, then the modified feasible location set might be R^2 with "holes" cut out; this causes S to become a nonconvex region and greatly complicates solution. In addition, firms may have to dispose of their toxic by-products at some specified disposal sites. Shipment to these sites needs to be included explicitly in the firm's production and location analysis (further complicating this issue is that shipment of these substances through certain regions or on certain edges of the transport network may be restricted). Especially if the site capacities are limited, such restrictions may reduced the optimal size and technology of new plants. Such regulatory restrictions can be handled, at least conceptually, by incorporating the pollutants in the production function and treating them similarly to other "inputs" that must be shipped (i.e., the pollutants are "disposal" inputs).

Because environmental regulations can impact firms' decisions through both the production and the locaton variables, an integrated production/location framework seems essential to understanding fully the consequences of such regulations. Yet, even the models presented here cannot capture all of the subtleties of industrial waste and pollution control. For example, in some localities the emission restrictions apply to the combined industry in a region; that is, as a group, the firms in a region, cannot produce more than L_i of some pollutant. Thus, the pollution generation of one plant affects the allowable level for other plants. Also some plants are both attractive and obnoxious in the sense that employees would like to live close to them for convenience but not too close because of their noise or pollution generation. For these reasons, researchers need to be looking toward even richer multiobjective production/investment/location models in the future.

Rate-of-Return Regulations

A type of regulation that has received great attention in the ecoomics literature is regulation of the rate of return that can be earned by a regulated firm. Regulated firms are generally those with some degree of monopoly power, such as electric utility companies. The seminal work on this topic was performed by Averch and Johnson [1962] who used a non-spatial model of a regulated monopoly under conditions of certainty to perform their analysis. They describe the reaction of profit-maximizing firms to regulatory constraints as follows: "if the rate of return allowed by the regulatory agency is greater than the cost of capital but is less than the rate of return that would be enjoyed by the firm were it free to maximize profit without regulatory constraint, then the firm will substitute capital for the other factor(s) of production and operate at an output where cost is not minimized" [Averch and Johnson 1962, p. 1053]. On the basis of this quotation Corey [1971, p. 358] concludes, "In short, the conventional regulatory constraints may induce regulated utilities to use too much capital vis a vis labor, fuel and other variable expenses and so not produce at the optimum social cost." In simple terms, rate-of-rturn regulation encourages more plant expansion or modernization than would otherwise be undertaken.

The maximum price regulated firms are allowed to charge their customers consists of costs associated with the operation of the firm. Included among those costs is the regulation-controlled maximum rate of return on capital. Part of the Averch-Johnson (A-J) proposition is that this rate of return on capital is greater than the cost of capital narrowly defined as the cost (e.g., interest) of obtaining the funds for the capital or stock inputs (but less than the return that could be earned by the firm if there were no regulatory constraints—that is the opportunity or "true" cost of capital). Thus, the effective price of the stock inputs utilized in the firm's calculations is lower than it would be without regulation.[2] Consequently there is a substitution in favor of stock input utilization. The relevant aspects of the A-J proposition for our purposes is that rate-of-return regulation may alter the firm's optimal input mix because it has the effect of artificially lowering the "price" of one of the inputs, capital. By inference, the optimal plant location may be affected as well. (The input mix and related locational implication of rate-of-return regulation, of course, depends on the appropriate technology supporting input substitutability.)

The only paper to consider the spatial implications of such regulation is by Mai [1985]. He extends the Averch-Johnson results to a model in which plant location is explicitly considered. Since regulation is often imposed to control a monopoly, Mai develops his model for a monopolist. The monopolist produces a single homogeneous output at rate z_0 using two inputs at "rates," z_1 and z_2 where z_1 is interpreted as labor and z_2 as capital. The production function is $z_0 = F(z_1, z_2)$ with $F_1 > 0, F_2 > 0, F_{11} < 0, F_{22} < 0$ (i.e., positive but decreasing marginal productivity). Notice that in this situation a dynamic treatment using stock-flow production functions would be preferable; i.e. where the stock of the capital input, rather than its flow of services, is the proper production function argument. Following tradition, and the approach in this book, Mai uses a static model and ignores the differences between stock and flow inputs. The use of stock-flow production functions in dynamic production/location models appears to be a reasonable direction for future research.

The location space is simplified to a line with the product market at one end, M_1, the labor market at the other, M_2, and capital is ubiquitous. (Note that this is opposite the convention used in (PL2.1); we will use Mai's convention to make comparison with his work easier.)[3] Once again, x is the location of the plant as well as the distance from M_1, and s is the distance from M_1 to M_2. The inverse demand function for product at market M_2 is $p_0 = p_0(z_0)$ with $p_0'(z_0) < 0$. The firm has no control over the prices, p_i, it pays for its inputs. The transportation costs per unit of product or labor transported are functions of the location, x, and are $r_0(x)$ and $r_1(x)$. This is a certainty model and the firm is taken to be a profit maximizer. Finally, the firm is prohibited from earning more than some fixed proportion, R, of the value of its (used) capital.

The problem for this firm is then to

$$\text{(PL8.1)}\quad \begin{aligned} \underset{z_1, z_2, x}{\text{maximize}} \quad & \pi = [p_0(z_0) - r_0(x)x]z_0 \\ & \quad - [p_1 + r_1(x)(s - x)]z_1 - p_2 z_2 & (8.1) \\ \text{subject to} \quad & \{[p_0(z_0) - r_0(x)x]z_0 & (8.2) \\ & \quad - [p_1 + r_1(x)(s - x)]z_1\}/z_2 \leqslant R \\ & z_0 = F(z_1, z_2) \end{aligned}$$

where p_2 is defined as the cost of a unit of capital on the location line. Thus, z_2 plays the role of the quantity of stock input employed and p_2 is the appropriate current cost of capital. The product, $p_2 z_2$, then is the annual cost of the stock input. Following the Averch-Johnson conditions, it is assumed that $R > p_2$. If $R = p_2$, the constraint restricts the firm to zero profit at best.

The Lagrangian function for (PL8.1) is

$$\begin{aligned} L = {} & [p_0(z_0) - r_0(x)x]F(z_1, z_2) - [p_1 + r_1(x)(s - x)]z_1 - p_2 z_2 \\ & - \lambda\{[p_0(z_0) - r_0(x)x]F(z_1, z_2) \\ & - [p_1 + r_1(x)(s - x)]z_1 - R z_2\} \end{aligned} \tag{8.3}$$

Assuming that the regulatory constraint is active, so that (8.2) is an equality, the first-order conditions can be written as follows:[4]

$$\begin{aligned} \partial L/\partial z_1 = {} & [p_0(\partial z_0/\partial z_1) + z_0(\partial p_0/\partial z_0)(\partial z_0/\partial z_1) \\ & - r_0(x)x(\partial z_0/\partial z_1)] - [p_1 + r_1(x)(s - x)] \\ & - \lambda[p_0(\partial z_0/\partial z_1) + z_0(\partial p_0/\partial z_0)(\partial z_0/\partial z_1) \\ & - r_0(x)x(\partial z_0/\partial z_1)] - [p_1 + r_1(x)(s - x)] \end{aligned}$$

$$= (1 - \lambda)\{[MR - r_0(x)x]F_1 - [p_1 + r_1(x)(s - x)]\} = 0 \quad (8.4a)$$

$$\partial L/\partial z_2 = (1 - \lambda)[MR - r_0(x)x]F_2 - p_2 + \lambda R = 0 \quad (8.4b)$$

$$\begin{aligned}\partial L/\partial x &= [-r_0(x) - x(dr_0(x)/dx)]F(z_1, z_2)\\ &\quad + [r_1(x) - (s - x)(dr_1(x)/dx)]z_1\\ &\quad - \lambda[-r_0(x) - x(dr_0(x)/dx)]F(z_1, z_2)\\ &\quad - \lambda[r_1(x) - (s - x)(dr_1(x)/dx)]z_1\\ &= -(1 - \lambda)[r_0(x) + x(dr_0(x)/dx)]F(z_1, z_2)\\ &\quad - (1 - \lambda)[-r_1(x) + (s - x)(dr_1(x)/dx)]z_1 = 0 \end{aligned} \quad (8.4c)$$

$$\begin{aligned}\partial L/\partial \lambda &= \{[p_0(z_0) - r_0(x)x]F(z_1, z_2)\\ &\quad - [p_1 + r_1(x)(s - x)]z_1 - Rz_2\}\end{aligned} \quad (8.4d)$$

where $MR = p_0 + z_0(\partial p_0/\partial z_0)$ is the marginal revenue.

Mai assumes an interior (nonendpoint) optimal location and that the second-order conditions are satisfied. In addition, he assumes that the second derivatives of the transport cost functions with respect to distance are zero; that is, $d^2r_0(x)/dx^2 = 0$ and $d^2r_1(x)/dx^2 = 0$. Furthermore he assumes that $2(dp_0/dz_0) + (d^2p_0/dz_0^2)z_0 < 0$; that is, $dMR/dz_0 < 0$ or marginal revenue decreases with increasing output.

Mai [1985, p. 457] uses the second-order conditions to establish the following.

Lemma

"If the firm maximizes its total profit subject to the production function and the regulatory constraint in which $R > p_2$ and if, in addition, the regulatory constraint is active and $z_1^* > 0$, $z_2^* > 0$, then we must have $0 < \lambda < 1$."

In order to investigate the impact of the regulation and to explore the Averch-Johnson results in this context we divide (8.4b) by (8.4a). Solving (8.4b) and (8.4a) for F_2 and F_1 respectively,

$$\begin{aligned}F_2 &= [p_2 - \lambda R]/\{(1 - \lambda)[MR - r_0(x)x]\}\\ &= \{p_2/[MR - r_0(x)x]\} - \{[\lambda(R - p_2)]/(1 - \lambda)[MR - r_0(x)x]\}\\ F_1 &= [p_1 + r_1(x)(s - x)]/[MR - r_0(x)x]\end{aligned}$$

Then

$$\frac{F_2}{F_1} = \frac{p_2}{p_1 + r_1(x)(s - x)} - \frac{\lambda(R - p_2)}{(1 - \lambda)[p_1 + r_1(x)(s - x)]} \tag{8.5}$$

Notice that $R > p_2$, and therefore from the lemma,

$$\frac{F_2}{F_1} < \frac{p_2}{p_1 + r_1(x)(s - x)} \tag{8.6}$$

Similar analysis of the first-order conditions for the *unconstrained* profit-maximizing problem shows that in that case

$$\frac{F_2}{F_1} = \frac{p_2}{p_1 + r_1(x)(s - x)} \tag{8.7}$$

So in the unconstrained case, as expected, the ratio of marginal productivities equals the ratio of the prices paid for the inputs. However, in the regulatory-constrained case the ratio of marginal productivities, with the marginal productivity of capital as the numerator, is less than the ratio of prices. With the assumption of diminishing marginal productivity this means that more capital and less labor is used in the optimal solution in the constrained case than in the unconstrained case or at least the capital-to-labor ratio is higher in the constrained case.

Next, consider the impact of regulation on the optimal location. The net locational pull (marginal transportation cost) given by equation (8.4c) can be written as

$$[r_0(x) + x(dr_0(x)/dx)]F(z_1, z_2) - [r_1(x) - (s - x)(dr_1(x)/dx)]z_1 = 0$$

(Again note that Mai is assuming the optimal location to be at an interior point on the line segment—an uncommon situation.) This is the same form the first-order condition resulting from $\partial L/\partial x$ would take without the regulatory constraint. For *given* levels of z_0 and z_1 the regulatory constraint has no impact on the optimal location. However, as we have already seen, the optimal values of z_0, z_1, and z_2 are influenced by the regulation; consequently, the optimal location is affected. Indeed, it would be expected that rate-of-return regulation, in the context of this model, would lead to a shift of location away from the labor market relative to the unconstrained (unregulated) case.

Mai also investigates the responses of z_0, z_1, z_2, and x to changes in the fair rate of return, R, by totally differentiating the first-order conditions. The analysis begins with R set at the unconstrained rate of return, R^*, then develops results as R is decreased to p_2. The analysis is straightforward but tedious and will not be reported in detail here. We will report some of the interesting results. For example:

Proposition

"The greater the difference between the regulatory fair rate of return R and the unconstrained profit-maximizing rate of return, R^*, the greater will be the capital input measured by z_2." Further: "The optimum location moves towards the site of the product market as the difference between the regulatory fair rate of return and the unconstrained profit-maximizing rate of return becomes greater, if capital and labor are complements (substitutes) and the marginal transportation cost is an increasing (decreasing) function of labor" [Mai 1985, pp. 459–460].

Thus, rate-of-return regulation does, potentially, have an impact on firms' location decisions and this impact should be taken into account, along with the distortion of input utilization and output rates, when rate-of-return regulation is being considered.

Mai's work, as well as much of the nonspatial regulatory research, assumed a deterministic environment. When uncertainty is introduced into the analysis, additional factors must be considered. For example, if the price of (cost) of fuel and/or labor are viewed as being more uncertain than the price of capital, the risk-averse firm would tend to use a more capital-intensive and less labor-intensive or fuel-intensive process than would be predicted from models that use the expected or average prices of each input. In this case, if an empirical study showed public utilities to be overly capital intensive (and many studies have done just this), it is not clear to what extent this is caused by price uncertainty and risk aversion or the A-J effect. Similarly, if the cost of capital is viewed as more risky than the prices of labor or fuel, risk aversion and the uncertainty would counteract the A-J effect. The importance of this "complication" has been discovered by Chapman and Waverman [1979; p. 108]: "This suggests that empirical work aimed at testing the Averch-Johnson effect should explicitly examine risk aversion. Otherwise the empirical results compound two capital intensity effects—risk aversion and regulation—without shedding light on the latter." Given the interesting results obtained from introducing uncertainty in chapters 5–7, stochastic production/location models under regulation seem to be worthy of future investigation.

Conclusion

In this chapter we have made use of our analytical framework to illustrate the effects of locational incentives and regulation. It was shown that spatial differentials in incentives are not necessary for incentives to affect firms' optimal locations; under uncertainty spatially uniform incentives can create spatial biases because of the input substitution and output effects caused by the uncertainty and risk preferences. We then showed how risk-reducing incentives could be designed to exploit the uncertainty and risk aversion of firms. Since firms are willing to pay to reduce the risks that they face, governments that share or absorb these risks for a premium can attract firms to their region and simultaneously generate additional revenue, as opposed to incurring the costs involved with other commonly used incentives. The risk-reducing action need not remove all risk from the firm nor must it necessarily be undertaken by governments. Private parties might be encouraged to provide risk-reducing options through the promotion of broad forms of business insurance, creation of futures markets, permitting premium pricing of utilities in exchange for uninterruptible service, and similar devices.

We also incorporated various forms of governmental regulation into the production/location framework. This exercise clearly demonstrated that regulations intended to be nonspatial in their purpose can, potentially, have significant spatial and production consequences. The production/location framework made identification of this phenonmenon possible, and therefore suggests that this framework be considered for more general policy analysis—even when the policy actions are *intended* to be nonspatial in their effects.

Notes

1. Suppose the price of input 1 changes over time, with the prices being realizations of $\tilde{p}_1$. Even though on occasion the realized price, p_1, might be greater than the guaranteed price, p_1^*, over the long term the government would pay an average price of $\bar{p}_1$ and sell at a price of p_1^*, thereby making an average profit of $p_1^* - \bar{p}_1$ per unit of input 1. If, on the other hand, the price of input 1 is set only once (being a realization of $\tilde{p}_1$), then the government could be operating at a loss forever if the realized price is greater than p_1^*. Even though the a priori expected profit for the government would be the $p_1^* - \bar{p}_1$, this latter situation would certainly be less appealing politically.

2. A more direct way of thinking about this situation is that the firm's profit is limited to a fixed rate, R, times capital; thus, it tries to maximize the amount of capital it uses subject to not adding capital that generates less than R return.

3. Mai claims that the assumption of capital being ubiquitous can be relaxed without substantively changing the results.

4. In (8.4c) Mai is assuming the optimal location to be at an interior point of the line segment. We have shown elsewhere that this is unlikely, and in many cases impossible. However, the results of his analysis are qualitatively still valid, and so we present his approach intact.

9 CONCLUSION

In this book we have attempted to provide a relatively complete and up-to-date presentation of integrated production/location theory. By incorporating the firm's input mix, output capacity, and facility location decisions within a single model we have been able to identify and illustrate how these decisions are interrelated. Analysis of these models made it clear how optimizing with respect to production variables alone—without regard to location—or optimizing with respect to location—without regard to the production variables—can lead to overall solutions that are suboptimal. In addition, the parametric analysis performed here demonstrated how changes in apparently nonspatial parameters, such as factor prices, exogenously set output levels, or spatially uniform government policies can have significant spatial effects on the firm.

Construction of optimization models for the typical firm that merge two traditionally distinct fields can provide significant conceptual insight into the firm's facility-planning problem. But there are other benefits as well. First, integrated production/location models may provide a better normative foundation for empirical studies of firm behavior. Specifically, using the production/location framework, we identified several microeconomic anomalies which could occur in practice and cause misleading empirical results. For example, in chapter 2 we found that even for well-behaved production functions, such as homogeneous functions, increases in output level could lead to decreases in the use of inputs that are not technologically inferior; this was due to a change in the optimal facility location which changed the delivered input prices. In chapter 5, we saw that parametric uncertainty and nonneutral risk preferences cause firms to use an input mix other than the cost-minimizing one which distorts (and can even create discontinuities in) the perceived cost function. Empirical derivations of production functions may incorrectly model the underlying production technology unless these production/location interactions are considered in the collection and analysis of the data.

Second, we have shown how in certain cases the more general production/location models reduce to relatively simpler, and well-known location models. In those cases the production/location problem can be solved more easily using well-established solution methods for the location models. In other cases it is possible to adapt solution methods developed for these simpler problems to solve the production/location problem with only slightly more effort than solving the location models. Thus, it is often possible to consider a firm's decision at a more general level rather than being forced to work immediately with a simplification of the problem. Consequently, the likelihood of obtaining a suboptimal solution for the firm's actual problem is reduced.

Third, it was shown in chapters 7 and 8 how integrated production/location models can be very useful for public policy evaluation. These models clearly demonstrated that even spatially uniform taxes, incentives, and regulations can have significant spatial effects on firms. This highlights the fact that governments should not focus entirely on spatially differential policies to attract firms. More importantly, policies that are intended to not affect facility location decisions of firms may still have a significant spatial effect; governments should consider this when instituting supposedly "nonspatial" policies. Our analysis of the models in chapters 7 and 8 also lead to the identification of possible strategies that governments might consider to help attract firms. Specifically, taxation policies and locational incentives can be designed in a way, at least in theory, so as to be mutually beneficial to both the government and to firms vis a vis traditional policies.

Research in the area of production/location theory has been substantial in the last 15 years, and it appears to be accelerating. However, much remains to be done. We conclude by listing and briefly discussing some of the areas in which additional work is merited; some of these are of a general nature, others are more narrow and technical.

1. The production/location models developed so far are all at a partial equilibrium (single firm) level. Although there is much that can be learned from partial equilibrium models, multifirm market or general equilibrium models are more valuable in studying the differential effects of parametric

changes. These are especially valuable when studying the effects of public policy actions, so that differential impacts and firm or market biases of policies can be identified. Of course, these models are more complex to analyze, which may be the reason they have received little study.

2. Many interesting enhancements can still be made to existing partial equilibrium models. By their nature, facility design and location decisions are long term and involve selection of "stock" inputs, such as capital equipment, which provide services over time, as well as "flow" inputs, such as fuel and raw materials. Generally, the amount and type of stock inputs selected at the plant construction stage restrict the flow-input-substitution options later. Thus, the firm's overall planning decision is dynamic in the sense that first an investment and location decision is made, and then later short-term production decisions are made. (Of course, at the first stage, anticipated short-term production plans are included in the decision.) This two-stage modeling of the production/location decision has only recently received attention. More general dynamic production/investment/location models are needed to capture the firm's actual decision problem.

3. Rapid technological change and relatively frequent geographic shifts in markets and sources of input supply have impacts on firms that are uncertain and are manifest over a period of time. The impacts involve redesign of plants, replacement of equipment, and relocation of facilities. The current pace of adoption of new technology in manufacturing and in information systems makes the interrelationships between technological obsolescence, equipment replacement, and plant relocation of substantial importance. Dynamic production/investment/location models may serve as a basis of analysis for this increasingly important phenomenon.

4. In recent years considerable production/location research has been directed toward models that include price uncertainty and general risk preferences. Other forms of uncertainty, such as technological, production, and transport rate uncertainty, merit study. This effort has already been initiated by Park and Mathur [1986a, 1986b] and Alperovich and Katz [1983], but more remains to be done, especially in the realm of parametric analysis. In fact, parametric analysis of stochastic production/location models appears to be one of the aspects most in need of study from both theoretical and applications points of view.

5. The need for multifirm models that include aspects of business uncertainty leads naturally to the issue of competitive location models. The only form of uncertainty studied by production/location researchers have been of an environmentally neutral nature. The assumption is that the values of one or more parameters, such as price or demand, are determined randomly, not selected by a third party so as to help or harm the

firm. This form of uncertainty is very common and of great interest to firms. Another business uncertainty that firms must consider is the behavior of competitors. Specifically, the demand for a firm's products or services may be influenced by the locations of competitors' facilities. (Sakashita [1980] and Hwang and Mai [1987] have considered vertically related firms in the production/location context. But in this case the firms are not competitors in the normal sense.)

The study of spatial competition and spatial equilibrium has a long history, dating back to at least Hotelling [1929]. Interest in spatial competition has been renewed in recent years (e.g., see Hansen and Thisse [1977], Wendell and McKelvey [1981], Hakimi [1983], Hurter and Lederer [1985], Lederer and Hurter [1986], and Dobson and Karmarkar [1987]). These models have generally assumed given cost functions and used relatively simple location spaces such as line segments (although more recent work has relaxed this). It appears that by allowing firms to select their technology (input mix) as well as location and, possibly, selling price, a richer set of competitive models would result. For example, such models might be able to suggest how types and degrees of competition would influence firms' technological as well as location choices. The existing models in spatial competition and production/location theory seem to provide a good foundation for integrative work on this topic.

6. The material in chapters 7 and 8 indicate that the production/location framework is potentially quite useful for evaluating public policies. This work, however, is still in its early stages. More extensive analysis of public policies within the existing production/location framework and especially within the enhanced models described above appears warranted. These models should aid public planners and decision makers to evaluate alternative policies and help in the development of new creative policies which can be used to promote and control the economic development of regions. For example, the possibility of tailoring tax policies and risk-reducing locational incentives to attract specific types of firms appears to be a promising use of such models.

7. Many computational problems remain in the study of production/location problems. For example, the modified Weiszfeld algorithm presented in chapter 4 appears to be potentially robust in its ability to solve production/location problems. The extent to which it can be guaranteed to converge for various location spaces, however, is an open question. This algorithm is a form of alternating algorithm, and it motivates the use of other alternating heuristics, such as the one described at the end of chapter 4. These solution methods appear promising, but very little empirical work has been published on the performance of solution algorithms for these problems.

It has taken over three years of writing and revising and several more years of preliminary thought and discussion to bring this book to fruition. In writing a book of this type there is always a problem with it being current and inclusive while not being tedious and dry. To improve readability we have tended to use the approach and emphasize the models with which we are most comfortable—those we helped develop through our research. The problem with such an approach is that it tends not to emphasize the extent and degree of contributions made by the numerous scholars in this field. The development of production/location theory has, in fact, been due to the work of many outstanding scholars and we do not intend to minimize their contributions. We thank all of those who have been instrumental in the development of this field, and apologize to anyone who feels their work received less attention than it deserved. We hope that the reader has found the book challenging, informative, and most of all, readable.

REFERENCES

Alonso, W. 1967. "A Reformulation of Classical Location Theory and Its Relation to Rent Theory." *Pap. Reg. Sci. Assoc.* 19:23–44.

Alperovich, G., and E. Katz. 1983. "Transport Rate Uncertainty and the Optimal Location of the Firm." *J. Reg. Sci.* 23:389–396.

Aly, A.A., and J. White. 1978. "Probabilistic Formulation of the Multifacility Weber Problem." *Naval Res. Logist. Quart.* 25:531–546.

Arrow, K.J. 1976. *Aspects of the Theory of Risk Bearing*. Jahnssonin Saatio. Helsinki.

Averch, H., and L.L. Johnson. 1962. "Behavior of the Firm Under Regulatory Constraint." *Am. Econ. Rev.* 52:1053–1069.

Balachandran, V., and S. Jain. 1976. "Optimal Facility Location Under Random Demand with General Cost Structure." *Naval Res. Logist. Quart.* 23:421–436.

Baron, D.P. 1970. "Price Uncertainty, Utility, and Industry Equilibrium in Price Competition." *Int. Econ. Rev.* 11:463–479.

Batra, R. 1974. "Resource Allocation in a General Equilibrium Model of Production Under Uncertainty." *J. Econ. Theory* 8:50–63.

Batra, R., and A. Ullah. 1974. "Resource Allocation in a General Equilibrium Model of Production Under Uncertainty." *J. Political Econ.* 82:537–548.

Berman, O., and M.R. Rahmana. 1985. "Optimal Location-Relocation Decisions on Stochastic Networks." *Transp. Sci.* 19:203–221.

Blair, R.D. 1974. "Random Input Prices and the Theory of the Firm." *J. Econ. Inquiry* 12:214–225

———. 1977. "Estimation of the Elasticity of Substitution When Input Prices are Random." *South. Econ. J.* 44:141–144.

Bradfield, M. 1971. "A Note on Location and the Theory of Production," *J. Reg.*

Sci. 11:263–266.

Brown, D.M. 1979. "The Location Decision of the Firm: An Overview of Theory and Evidence," *Pap. Reg. Sci. Assoc.* 43:23–39.

Carbone, R., and A. Mehrez, 1980. "The Single Facility Minimax Distance Problem Under Stochastic Location of Demand." *Manage. Sci.* 26:113–115.

Chapman, R., and L. Waverman. 1979. "Risk Aversion, Uncertain Demand, and the Effects of a Regulatory Constraint." *J. Public Econ.* 11:107–121.

Chiu, S.S., and R. Larson, 1985. "Locating an N-Server Facility in a Stochastic Environment." *Comp. Oper. Res.* 12:509–516.

Clapp, J.M. 1983. "A General Model of Equilibrium Locations." *J. Reg. Sci.* 23: 461–478.

Clarke, H.R. 1984. "The Separability of Production and Location Decisions: Comment." *Am. Econ. Rev.* 74:528–530.

Clarke, H.R., and R.M. Shrestha. 1983. "Location and Input Mix Decisions for Energy Facilities." *Reg. Sci. Urban Econ. 13*:487–504.

Coes, D.V. 1977. "Firm Output and Changes in Uncertainty." *Am. Econ. Rev.* 67:249–251.

Collins, L., and D.F. Walker (eds.). 1975. *Locational Dynamics of Manufacturing Activity.* Wiley: New York.

Cooper, L. 1963. "Location-Allocation Problems." *Oper. Res.* 11:331–343.

———. 1964. "Heuristic Methods for Location-Allocation Problems." *SIAM Rev.* 6:37–52.

———. 1967. "Solutions of Generalized Locational Equilibrium Problems." *J. Reg. Sci.* 7:1–18.

———. 1974. "A Random Locational Equilibrium Problem." *J. Reg. Sci.* 14: 47–54.

———. 1978. "Bounds on the Weber Problem Solution Under Conditions of Uncertainty." *J. Reg. Sci.* 18:87–93.

Corey, G. 1971. "The Averch and Johnson Proposition: A Critical Analysis." *Bell J. Econ.* 2:358–373.

Corneujols, G., M. Fisher, and G. Nemhauser. 1977. "Location of Bank Accounts to Optimize Float: An Analytic Study of Exact and Approximate Algorithms." *Manage. Sci.* 23:789–810.

Diewert, W.E. 1974. "Applications of Duality Theory." In M.D. Intriligator and D.A. Kendrick (eds.), *Frontiers of Quantitative Economics. Volume II.* North Holland: Amsterdam.

Dobson. G., and U.S. Karmarkar. 1987. "Competitive Location on a Network." *Oper. Res.* 35:565–574.

Domar, E.D., and R.A. Musgrave. 1944. "Proportional Income Taxation and Risk Taking." *Quart. J. Econ.* 58:388–442.

Drezner, Z. 1979. "Bounds on the Optimal Location to the Weber Problem Under Conditions of Uncertainty." *J. Oper. Res. Soc.* 30:923–931.

Drezner, Z., and G.O. Wesolowsky. 1981. "Optimum Location Probabilities in the L_P Distance Weber Problem." *Transp. Sci.* 15:85–97.

Duffin, R.J., E.L. Peterson, and C. Zener. 1967. *Geometric Programming—*

Theory and Applications. Wiley: New York.

Efroymson, M.A., and T.L. Ray. 1966. "A Branch-Bound Algorithm for Plant Location." *Oper. Res.* 14:361–368.

Emerson, D.L. 1973. "Optimum Firm Location and the Theory of Production." *J. Reg. Sci.* 13:335–347.

Epstein, L. 1978. "Production Flexibility and the Behavior of the Competitive Firm Under Price Uncertainty." *Rev. Econ. Stud.* 45:251–261.

Erlenkotter, D. 1978. "A Dual-based Procedure for Uncapacitated Facility Location." *Oper. Res. 26*:992–1009.

Eswaran, M., Y. Kanemoto, and D. Ryan [EKR]. 1981. "A Dual Approach to the Locational Decision of the Firm." *J. Reg. Sci.* 21:469–490.

Eyster, J.W., J.A. White, and W.W. Wierwille. 1973. "On Solving Multifacility Location Problems Using a Hyperboloid Approximation Procedure." *AIIE Trans.* 5:1–6.

Feldstein, M.S. 1971. "Production with Uncertain Techology: Some Economic and Econometric Implications." *Int. Econ. Rev.* 12:27–37.

Ferguson, C.E., and T.R. Saving. 1969. "Long-Run Scale Adjustments of a Perfectly Competitive Firm and Industry." *Am. Econ. Rev.* 59:774–783.

Fishburn, P.C. 1977. "Mean-Risk Analysis with Risk Associated with Below-Target Returns." *Am. Econ. Rev.* 67:116–126.

Fox, W.F. 1981. "Fiscal Differentials and Industrial Location: Some Empirical Evidence." *Urban Stud.* 18:105–111.

Franca, P.M., and H.P.L. Luna. 1982. "Solving Stochastic Transportation-Location Problems by Generalized Benders Decomposition." *Transp. Sci.* 16:113–126

Francis, R.L., and A.V. Cabot. 1972. "Properties of a Multifacility Location Problem Involving Euclidean Distances." *Naval Res. Logist. Quart.* 19:335–353.

Francis, R.L., L.F. McGinnis, and J.A. White. 1983. "Locational Analysis." *Eur. J. Oper. Res. 12*:220–252.

Francis, R.L., and J.A. White. 1974. *Facility Layout and Location: An Analytical Approach*. Prentice-Hall: Englewood Cliffs, N.J.

Friedman, M., and L.J. Savage. 1948. "The Utility Analysis of Choices Involving Risk." *J. Political Econ.* 56:279–304.

Geoffrion, A.M., and R.D. McBride. 1978. "Lagrangean Relaxation Applied to Capacitated Facility Location Problem." *AIIE Trans.* 10:40–47.

Ghosh, A., and G. Rushton. 1987. *Spatial Analysis and Location-Allocation Models*. van Nostrand Reinhold: New York.

Goldman, A.J. 1974. "Fixed Point Solution of Plant Input/Location Problems." *J. Res. Nat. Bur. Stand.* 78B:79–94.

Hakimi, S.L. 1964. "Optimal Locations of Switching Centers and the Absolute Centers and Medians of a Graph." *Oper. Res.* 12:405–459.

———. 1983. "On Locating New Facilities in a Competitive Environment." *Eur. J. Oper. Res.* 12:29–35.

Handler, G., and P.B. Mirchandani. 1979. *Location on Networks*. M.I.T. Press:

Cambridge, MA.

Hansen, P., and J.-F. Thisse. 1977. "Multiplant Location for Profit Maximization." *Environ. Plan. A* 9:63–73.

Hartman, R. 1976. "Factor Demand with Output Price Uncertainty." *Am. Econ. Rev.* 66:675–681.

Heaps. T. 1982. "Location and the Comparative Statics of the Theory of Production." *J. Econ. Theor.* 28:102–112.

Higano, Y. 1985. "On the 'Exclusion Theorem'," *Reg. Sci. Urban Econ.* 15: 449–458.

Hodder, J.E., and J.V. Jucker. 1985. "A Simple Plant-Location Model for Quantity-Setting Firm Subject to Price Uncertainty." *Eur. J. Oper. Res.* 21:36–49.

Hoover, E.M. 1948. *The Location of Economic Activity.* McGraw-Hill: New York.

Horowitz, I. 1970. *Decision Making and the Theory of the Firm.* Holt, Rinehart and Winston: New York.

Hotelling, H. 1929. "Stability in Competition." *Econ. J.* 39:41–57.

Hsu., S., and C. Mai. 1984. "Production Location and Random Input Price." *Reg. Sci. Urban Econ.* 14:45–62.

Hurter, Jr., A.P., and P.J. Lederer. 1985. "Spatial Duopoly with Discriminatory Pricing," *Reg. Sci. Urban Econ.* 15:541–553.

Hurter, Jr., A.P., and J.S. Martinich. 1984a. "Network Production-Location Problems Under Price Uncertainty." *Eur. J. Oper. Res.* 16: 183–197.

———. 1984b. "Firm Size and Risk Characteristics and the Production-Location Problem Under Factor Price Uncertainty." Working Paper, Dept. of Industrial Engineering and Management Sciences, Northwestern University, Evanston, Ill.

———. 1985. "Input Price Uncertainty and the Production-Location Decision: A Critique and Synthesis." *Reg. Sci. Urban Econ.* 15:591–596.

———. 1987. "Production-Location Decisions Under Uncertainty: Optimization Models and Policy Analysis." In *Models in Urban Geography 4B.* Concept Publishing Company: New Delhi.

Hurter, Jr., A.P., J.S. Martinich, and E.R. Venta [HMV]. 1980. "A Note on the Separability of Production and Location." *Am. Econ. Rev.* 70:1042–1045.

———. 1984. "The Separability of Production and Location: Reply." *Am. Econ. Rev. 1984.* 74:531–532.

Hurter, Jr., A.P., and M.A. Moses. 1967. "Price and Productive Uncertainties in Dynamic Planning." *J. Reg. Sci.* 7:33–47.

Hurter, Jr., A.P., and J. Prawda. 1970. "A Warehouse Location Problem with Probabilistic Demand." Working Paper Series No. 42, Graduate School of Business Administration, Tulane University.

Hurter, Jr., A.P., and E.R. Venta. 1982. "Production-Location Problems." *Naval Res. Logist. Quart.* 29:279–290.

Hurter, Jr., A.P. and R.E. Wendell. 1972. "Location and Production—A Special Case." *J. Reg. Sci.* 12:243–247.

Hwang, H. and C-C. Mai. 1987. "Industrial Location and Rising Energy Prices: A Case of Bilateral Monoply," *Reg. Sci. Urban Econ.* 17:255–264.

Isard, W. 1956. *Location and Space Economy*. Wiley: New York.

Jucker, J.V., and R.C. Carlson, 1976. "The Simple Plant-Location Problem Under Uncertainty." *Oper Res.* 24:1045–1055.

Juel, H. 1980. "A Note on Bounds on the Weber Problem Solution Under Conditions of Uncertainty." *J. Reg. Sci.* 20:523–524 (with Cooper's reply).

———. 1981. "Bounds in the Generalized Weber Problem Under Uncertainty." *Oper. Res* 29:1219–1227.

Katz, E. 1984. "The Optimal Location of the Competitive Firm Under Price Uncertainty." *J. Urban. Econ.* 16:65–75.

Katz, I.N. 1969. "On the Convergence of a Numerical Scheme for Solving Some Location Equilibrium Problems." *SIAM J. Appl. Math.* 17:1224–1231.

Katz, I.N., and L. Cooper. 1974. "An Always-Convergent Numerical Scheme for a Random Locational Equilibrium Problem." *SIAM J. Num. Anal.* 11:683–692.

———. 1976. "Optimal Facility Location for Normally and Exponentially Distributed Points." *J. Res. Nat. Bur. Stand.* 80B:53–73.

Keeney, R.L., and H. Raiffa. 1976. *Decisions with Multiple Objectives*. Wiley: New York.

Khalili, A., V.K. Mathur, and D. Bodenhorn [KMB]. 1974. "Location and the Theory of Production: A Generalization." *J. Econ. Theor.* 9:467–475.

Khumawala, B.M. 1972. "An Efficient Branch-and-Bound Algorithm for the Warehouse Location Problem." *Manage. Sci.* 18:718–731.

Krarup, J., and P.M. Pruzan. 1983. "The Simple Plant Location Problem: Survey and Synthesis." *Eur. J. Oper. Res.* 12:36–81.

Kuhn, H.W. and R.E. Kuenne. 1962. "An Efficient Algorithm for the Numerical Solution of the Generalized Weber Problem in Spatial Economics." *J. Reg. Sci.* 4:21–33.

Kusumoto, S-I. 1984. "On a Foundation of the Economic Theory of Location-Transport Distance vs. Technological Substitution." *J. Reg. Sci.* 24:249–270.

Laridon, P. 1984. "A Dual Approach to the Genralized Weber Problem Under Locational Uncertainty." *CCERO (Belgium)* 26:241–253.

Lawler, E.L. 1976. *Combinatorial Optimization: Networks and Matroids*. Holt, Rinehart and Winston: New York.

Lederer, P.J., and A.P. Hurter, Jr. 1986. "Competition of Firms: Discriminatory Pricing and Location." *Econometrica* 54:623–640.

Leland, H. 1972. "Theory of the Firm Facing Uncertain Demand." *Am. Econ. Rev.* 62:278–291.

Levy, J. 1967. "An Extended Theorem for Location on a Network." *Oper. Res. Quart.* 18:433–442.

Lippmann, S., and J.J. McCall. 1982. "Taxation, Incentives and Risk-Sharing." *Oper. Res. Letters* 1:83–84.

Louveaux, F., and J-F. Thisse. 1985. "Production and Location on a Network Under Demand Uncertainty. *Oper. Res. Let.* 4:145–149.

Love, R.F., J.G. Morris, and G.O. Wesolowsky. 1988. *Facilities Location: Models and Methods*. North-Holland: New York.

Luce, R.D. and H. Raiffa. 1957. *Games and Decisions*. Wiley: New York.

Mai, C. 1981. "Optimum Location and the Theory of the Firm Under Demand Uncertainty. *Reg. Sci. Urban Econ.* 11:549–557.

———. 1984a. "Demand Function and Location Theory of the Firm Under Price Uncertainty." *Urban Stud.* 21:459–464.

———. 1984b. "Location and the Theory of the Imperfectly Competitive Firm Under Demand Uncertainty." *South. Econ. J.* 50:1160–1170.

———. 1985. "Optimum Location and Theory of the Firm Under a Regulatory Constraint." *J. Reg. Sci.* 25:453–461.

Martinich, J.S. 1980. *Production-Location Problems Under Price Uncertainty*. Unpublished Ph.D. Dissertation, Northwestern University, Evanston, Ill.

Martinich, J.S., and A.P. Hurter, Jr. 1982. "Price Uncertainty and the Optimal Production-Location Decision." *Reg. Sci. Urban Econ.* 12:509–528.

———. 1985. "Price Uncertainty, Factor Substitution, and the Locational Bias of Business Taxes." *J. Reg. Sci.* 25:175–190.

———. 1987. "A Note on Income Taxes and Degree One Production Homogeneity." *J. Reg. Sci.* 27:477–482.

Martinich, J.S. and A.P. Hurter., Jr. 1988. "Generalized Comparative Statics for the Production-Location Problem," Working Paper, School of Business Administration, University of Missouri-St. Louis.

Mathur, V.K. 1979. "Some Unresolved Issues in the Location Theory of the Firm." *J. Urban Econ.* 6:299–318.

———. 1983. "Location Theory of the Firm Under Price Uncertainty." *Reg. Sci. Unban Econ.* 13:411–428

———. 1985. "Location Theory of the Firm Under Price Uncertainty: Some New Conclusions." *Reg. Sci. Urban Econ.* 15:597–598.

Mayshar, J. 1977. "Should Governments Subsidize Risky Private Projects?" *Am. Econ. Rev.* 67:20–28.

Miller, S.M., and O.W. Jensen. 1978. "Location and the Theory of Production." *Reg. Sci. Urban Econ.* 8:117–128.

Minieka. E. 1977. "The Centers and Medians of a Graph." *Oper. Res.* 25:641–650.

Mirchandani, P.B. 1980. "Location Decision on Stochastic Networks." *Geogr. Anal.* 12:172–183.

Mirchandani, P.B., and A.R. Odoni. 1979. "Locations of Medians on Stochastic Networks." *Trans. Sci.* 13:85–97.

Mirchandani, P.B., and A. Oudjit. 1980. "Localizing 2-Medians on Probabilistic and Deterministic Tree Networks." *Networks* 10:329–350.

Morris, J.G. 1981. "Convergence of the Weiszfeld Algorithm for Weber Problems Using a Generalized "Distance" Function," *Oper Res.* 29:37–48.

Morris, J.G., and W.A. Verdini. 1979. "Minisum L_p Distance Location Problems Solved via a Perturbed Problem and Weiszfeld's Algorithm." *Oper. Res.* 27: 1180–1188.

Moses, L.N. 1958. "Location and the Theory of Production." *Quart. J. Econ.* 72:259–272.

Mossin, J. 1968. "Taxation and Risk-Taking: An Expected Utility Approach." *Economica* 35:74–82.

Nauss, R.M. 1978. "An Improved Algorithm for the Capacitated Facility Location Problem." *J. Oper. Res. Soc.* 29:1195–1201.

Nijkamp, P., and J. Paelink. 1973. "A Solution Method for Neo-Classical Location Problems." *Reg. Sci. Urban Econ.* 3:383–410.

Park, K., and V.K. Mathur. 1986a. "Location Theory of the Firm Under Uncertainty: A generalization." Unpublished Manuscript, Cleveland State University, 29 pages.

———. 1986b. "Production-Technology Uncertainty and the Optimal Location of the Firm." Unpublished Manscript, Cleveland State University (24 pages).

Pindyck, R.S. 1982. "Adjustment Costs, Uncertainty, and the Behavior of the Firm." Am. Econ. Rev. 72:415–427.

Planchart, A., and A.P. Hurter, Jr., 1975. "An Efficient Algorithm for the Solution of the Weber Problem with Mixed Norms." *SIAM J. Contol* 13:650–665.

Pope, R.D., and R.E. Just. 1978. Uncertainty in Production and the Competitive Firm: Comment." *South. Econ. J.* 44:669–674.

Pope, R.D., and R.A. Kramer. 1979. "Production Uncertainty and Factor Demands for the Competitive Firm." *South. Econ. J.* 46:489–501.

Pratt, J.W. 1964. "Risk Aversion in the Small and in the Large," *Econometrica* 32:122–136.

Predohl, A. 1928. "The Theory of Location in its Relation to General Economics," *J. Political Econ.* 36:371–390.

Puu, T. 1971. "Some Comments on 'Inferior' (Regressive) Inputs." *Swedish J. Econ.* 73:241–251.

Raiffa, H. 1968. *Decision Analysis*. Addison-Wesley: Reading, Mass.

Ratti, R.A. 1978. "Uncertainty in Production and the Competitive Firm: Reply." *South. Econ. J.* 45:675–679.

———, and A. Ullah. 1976. "Uncertainty in Production and the Competitive Firm." *South. Econ. J.* 43:703–710.

Rech, P., and L.G. Barton. 1970. "A Non-convex Transportation Algorithm." In E.M.L. Beale (ed.), *Applications of Mathematical Programming Techniques*. American Elsevier: New York.

Rothschild, M., and J.E. Stiglitz. 1970. "Increasing Risk I: A Definition." *J. Econ. Theor.* 2:225–243.

Sakashita, N. 1967. "Production Function, Demand Function and Location Theory of the Firm." *Pap. Reg. Sci. Assoc.* 20:109–122.

Sakashita, N. 1980. "The Location Theory of Firm Revisited: Impacts of Rising Energy Prices," *Reg. Sci. Urban Econ.* 10:423–428.

Samuelson, P.A. 1976. *Foundations of Economic Analysis*. Atheneum: New York.

Sandmo, A. 1971. "Competitive Firm Under Price Uncertainty." *Am. Econ. Rev.* 61:65–73.

Savage, L.J. 1954. *The Foundations of Statistics*. Wiley: New York.

Schmenner, R.W. 1982. *Making Business Location Decisions*. Prentice-Hall: Englewood Cliffs, N.J.

Schrage, L. 1975. "Implicit Representation of Variable Upper Bounds in Linear Programming." *Math. Programming Study* 4:118–132.

Scott, A.J. 1971. *Combinatorial Programming, Spatial Analysis, and Planning*. Methuen: London.

Seppala, Y. 1975. "On a Stochastic Multi-Facility Location Problem." *AIIE Trans.* 7:56–62.

Shephard, R.W. 1970. *Theory of Cost and Production Functions*. Princeton University Press: Princeton, N.J.

Shieh, Y-N. 1983a. "The Moses-Predohl Pull and the Neoclassical Location Theory. *Reg. Sci. Urban Econ.* 13:517–524.

Shieh, Y-N. 1983b. "The Space Cost Curve and Variable Transport Costs: A General Production Function Case," *Urban Studies* 20:241–245.

———. 1985. "A Note on the Clarke and Shrestha Linear Space Model," *Reg. Sci. Urban Econ.* 15:131–135.

———. 1987. "Increasing Returns to Scale and Location Theory of the Firm Under Price Uncertainty," *Urban Studies* 24:163–166.

Shieh, Y-N., and C. Mai. 1984. "Location and the Theory of Production." *Reg. Sci. Urban Econ.* 14:199–218.

Soland, R.M. 1974. "Optimal Plant Location with Concave Costs." *Oper. Res.* 22:373–385.

Stafford, H.A. 1980. *Principles of Industrial Facility Location*. Conway Publications: Atlanta, GA.

Stewart, M.B. 1978. "Factor-Price Uncertainty with Variable Proportions." *Am. Econ. Rev.* 68:468–473.

———. 1982. "Target Returns and the Theory of the Firm Under Uncertainty." *Zeitschrift fur Nationalokonomie* 42:143–152.

Thisse, J.-F., and J. Perreur. 1977. "Relations Between the Point of Maximum Profit and the Point of Minimum Total Transportation Cost: A Restatement." *J. Reg. Sci.* 17:227–234.

Turnovsky, S. 1973. "Production Flexibility, Price Uncertainty and the Behavior of the Competitive Firm." *Int. Econ. Rev.* 14:395–413.

van Roy, T.J. 1986. "A Cross Decomposition Algorithm for Capacitated Facility Location." *Oper. Res.* 34:145–163.

Venta, E.R., and A.P. Hurter, Jr. 1985. "Production-Location Problems with Demand Considerations." *Naval Res. Logist. Quart.* 32:625–630.

von Neumann, J., and O. Morgenstern. 1947. *Theory of Games and Economic Behaviour*. 2nd Ed., Princeton University Press: Princeton, N.J.

Weaver, J.R., and R.L., Church. 1983. "Computational Procedures for Location Problems on Stochastic Networks." *Transp. Sci.* 17:168–190.

Weiszfeld, E. 1937. "Sur le point pour lequella somme des distances de n points donnes est minimum." *Tohoku Math. J.* 43:355–386.

Wendell, R. E., and A. P. Hurter, Jr. 1973a. "Location Theory, Dominance and Convexity." *Oper. Res.* 21:314–319.

———. 1973b. "Optimal Locations on a Network." *Transp. Sci.* 7:18–33.

———. 1976. "Minimization of a Non-Separable Objective Function Subject to Disjoint Constraints." *Oper. Res. 24*:643–657.

Wendell, R. E., and R. D. McKelvey. 1981. "New Perspectives in Competitive Location Theory." *Eur. J. Oper. Res.* 6:174–182.

Wendell, R. E., and E. L. Peterson. 1984. "A Dual Approach for Obtaining Lower Bounds to the Weber Problem." *J. Reg. Sci.* 24:219–228.

Wendell, R. E., and D. M. Rosenblum. 1980. "Further Results on the Minimization of a Nonseparable Objective Function Subject to Disjoint Constraints." *Oper. Res.* 28:1222–1226.

Wesolowsky, G. O. 1977. "The Weber Problem with Rectangular Distances and Randomly Distributed Destinations." *J. Reg. Sci.* 17:53–60.

Wesolowsky, G. O., and R. F. Love. 1972. "A Nonlinear Approximation Method for Solving a Generalized Rectangular Distance Weber Problem." *Manage. Sci.* 18:656–663.

Woodward, R. S. 1973. "The Iso-Outlay Function and Variable Transport Costs." *J. Reg. Sci.* 13:349–355.

Ziegler, J. A. 1986. "Location, Theory of Production, and Variable Transportation Rates." *J. Reg. Sci.* 26:785–791.

Author Index

Subject Index